Rheology of Fresh
Cement and Concrete

Publisher's Note
This book has been produced from camera ready copy provided
by the individual contributors whose cooperation is gratefully
acknowledged. This has facilitated rapid publication of the
Proceedings.

Rheology of Fresh Cement and Concrete

Proceedings of the International Conference
organized by the British Society of Rheology,
University of Liverpool, UK
March 16–29 1990

Edited by

P.F.G. BANFILL

CRC Press
Taylor & Francis Group
Boca Raton London New York

CRC Press is an imprint of the
Taylor & Francis Group, an **informa** business
A TAYLOR & FRANCIS BOOK

CRC Press
Taylor & Francis Group
6000 Broken Sound Parkway NW, Suite 300
Boca Raton, FL 33487-2742

First issued in paperback 2019

© 1991 by Taylor & Francis Group, LLC
CRC Press is an imprint of Taylor & Francis Group, an Informa business

No claim to original U.S. Government works

ISBN-13: 978-0-419-15360-3 (hbk)
ISBN-13: 978-0-367-86364-7 (pbk)
ISBN-13: 978-0-442-31254-1 (USA)

British Library Cataloguing in Publication Data
Available

Library of Congress Cataloging-in-Publication Data
Available

Publisher's Note
The publisher has gone to great lengths to ensure the quality of this reprint but points out that some imperfections in the original may be apparent

Visit the Taylor & Francis Web site at
http://www.taylorandfrancis.com

and the CRC Press Web site at
http://www.crcpress.com

Contents

vi

Preface

This book contains the papers presented at the International Conference on Rheology of Fresh Cement and Concrete held at the University of Liverpool in March 1990. Although the subject has featured as sub-themes at other conferences of broader scope, I believe that this is the first conference devoted entirely to the rheology of cement and concrete, indicating that the subject has reached maturity.

The rheology of cement and concrete governs the ease with which these materials flow and is influenced by composition and a variety of external factors. It can be argued that rheology is critically important for all uses of cement and concrete since the desirable hardened properties of strength and durability can be achieved only if the fresh material is capable of flowing into, and being compacted, to fill moulds or other interstices. This importance is now beginning to be recognized by engineers and other users, but because of the complexity of the properties of these materials some of the answers to industrial problems take a long time to emerge. Nevertheless, as these papers show, rheological measurements can give much information that is of use in such industrially important areas as oilwell cementing, grouting, vibratory compaction and quality control. In general, theoretical modelling has lagged behind phenomenological studies of rheology, but the papers in this area show that predictions about the behaviour of cement systems may soon be possible.

The papers in this book are grouped into sections dealing with different types of materials – cement pastes in general, oilwell cement slurries, grouts and concretes. In addition, there are sections dealing with the influence of vibration on cement pastes and concretes and theoretical studies. The final section contains discussion contributions on the papers.

The preparation of this book has involved many people. I would like to thank all the authors for preparing their papers to a high standard and to schedule. I would like to acknowledge especially the help of Dr T. Gregory and Dr G.H. Tattersall, my colleagues on the Technical Committee, and Nick Clarke, the publisher. Finally, I appreciated the wise help of fellow members of the Council of the British Society of Rheology.

P.F.G. Banfill
Liverpool
May 1990

PART ONE

CEMENT PASTES: EFFECTS OF CHEMICAL COMPOSITION, ADMIXTURES AND LATENT HYDRAULIC BINDERS

1 RHEOLOGICAL MEASUREMENTS WITH CEMENT PASTES IN VISCOMETERS: A COMPREHENSIVE APPROACH

J. CHAPPUIS
Lafarge Coppée Recherche, Viviers/Rhone, France

Abstract
Because of the small size of cement particles, interparticle forces play a very important role. The rheology of cement pastes is highly related to the state of flocculation of the cement particles. It is then necessary to refer to Colloïd Science to have a better understanding of the rheology of fresh cement pastes and concrete.
When attractive interparticle forces dominate, there is in the paste an internal structure which is responsible for poor flow properties. This internal structure can be destroyed by agitation and the paste is then fluidified (thixotropy).
In mortar or concrete, the cement paste is submitted to very small shear stresses during the pouring of the material into a mould or formwork. In order to simulate the shearing conditions during concrete placing, rheological measurements in viscometers must be carried at very slow speeds and during very short times. From such considerations, we have developped an appropriate procedure to study the rheology of cement pastes as a function of time during the dormant period.
Experimental results obtained with such a procedure are presented and discussed for three types of systems:
- model dense suspensions (alumina particles in water, either in a flocculated or in a non-flocculated state)
- aluminous cement pastes made from normal and stiffening cements
- portland cement pastes at different water to cement ratios
Keywords: Rheological measurements, Flocculation, Interparticle forces, Thixotropy, Alumina suspensions, Aluminous cement, Portland cement, Stiffening

1 General background of the rheology of fresh cement and concrete

Concrete is a mixture of aggregates, cement and water. Cement particles range between about 1 μm and 100 μm. Aggregates range from a fraction of a millimeter to a few centimeters. The constituents of concrete are energetically blended for a short time in a mixer (a few minutes or less) in order to get a uniform mass. After this mixing, we can consider that the aggregates are embedded in the cement paste, which is a dense suspension of cement particles in water.
After an induction period called "dormant period" there are chemical reactions (cement + water --> hydrates), which lead to the hardening of the cement paste and finally to the obtention of an artificial stone. The rheological properties of fresh concrete are practically important at the time when it is placed in formworks and that can be done at any time during the dormant period.
There is a large number of empirical tests for measuring the rheological properties of fresh concrete or mortar. For example: slump, flow test, V-B test, maniabilimètre

L.C.L., ball penetration test or the two-point test as proposed by Tattersall et al. (1979); these tests are fully described in the literature, for instance by J. Baron (1982).

In most tests, the shear rate is variable in the mass of the tested concrete, and at each point, it can also vary with time. As concrete is a non-Newtonian fluid, and as its flow properties are influenced by the amount of shear it experiences, the correlation between the different workability tests is poor.

Fresh concrete and fresh mortar contain particles which are too coarse for conventional laboratory viscometers, in which all the tested material is sheared at the same rate, in the narrow space between two coaxial cylinders or cones. It is then not possible to obtain pure scientific quantities describing the rheology of these materials.

Since concrete is a suspension of inert aggregates in a matrix of cement paste, the workability of concrete depends on parameters related to the aggregates (quantity, size distribution, shape...), but it is also highly related to the rheology of the cement paste which constitutes that matrix. A large number of studies have been carried on cement pastes with coaxial cylinders viscometers: for instance in the case of Portland cement by Roy et al. (1979,1980), Banfill et al. (1981), Lapasin et al. (1983), Atzeni et al. (1983), and in the case of aluminous cement by Banfill et al. (1986) and Chappuis et al. (1988, 1989). In a review paper on this subject, Costa and Massazza (1986) conclude that the understanding of the phenomena concerning the paste rheology has not yet reached a sufficient level to interpret the more complex "suspension" formed by fresh concrete.

There are several parameters during the process of manufacturing and ageing of the cement which can have important effects on the rheology of concrete. Depending of these parameters the result can be a stiffening of the concrete and this phenomenon is then well correlated with an abnormal rheological behaviour of the cement paste.

For studying the effect of cements on the rheology of concrete, we think that scientific measurements on cement pastes are preferable to empirical tests carried on concretes, even in the presence of additives.

2 The important effect of interparticle forces on the rheology of cement suspensions

A cement paste is a dense suspension of cement particles in water. As the mean size of the cement particles is on the order of 10 micrometers, interparticle forces and gravity are of the same order of magnitude and both types of forces play an important role concerning the macroscopic characteristics of such suspensions. As mentioned by Adamson (1982) for instance, interparticle forces are of two kinds: van der Waals forces and electrical double layer interactions.

In a system made of two particles there is an attractive London - van der Waals force between every couple of atoms of the system. For two particles of radius R, separated by a distance d, the resulting macroscopic van des Waals force F can be expressed, when $d \ll R$, by $F = HR/12\ d^2$, H being the Hamaker coefficient of the system. Hamaker (1937) has established that in a liquid medium, when two particles are made of the same material, the Hamaker coefficient is positive and the van der Waals force is always attractive, whatever the liquid.

The electrical double layer interactions have their origin in the surface electric charge which appears for most substances in contact with an aqueous medium; for instance, with regards to inorganic oxides, Furlong et al. (1981) have established that the electrical double layer at the oxide/water interface is related to the unequal adsorption of H^+ and OH^-. The electric double layer at the surface of small particles can be experimentally investigated by electrophoretic measurements. Between two particles the double layer interaction can be attractive or repulsive according to the sign of the charge of each of the particles.

It is experimentally known and it has been theoretically demonstrated in the D.L.V.O. theory established by Derjaguin and Landau (1941) and by Verwey and Overbeek (1948), that suspensions are stable when repulsive double layer interactions are larger than van der Waals forces and that there is flocculation in the opposite case. Regarding the rheological properties, and as stated by Chappuis (1982, 1986), suspensions have good flowing properties when repulsive interparticle forces dominate, that is to say when the suspension is deflocculated. When repulsive double layer interactions are smaller than attractive van der Waals forces, the solid particles form a structure inside the liquid and the suspension is very cohesive and has poor flow properties.

3 Fluidification of a cement paste under agitation (thixotropy)

Let us consider a very simple experiment: a stiffening aluminous cement has been specially manufactured for this study in a way that the concretes made with that cement exhibit an important stiffening during the dormant period. With this cement, we have prepared a paste with a water to cement ratio of 0.5 and this paste has been divided into two portions in two different beakers. The first beaker is periodically agitated by a manual moderate shaking with a spatula and even one hour later the paste has the consistency of a fluid liquid. In the second beaker, the surface of the paste is from time to time smoothly touched with the tip of the spatula and after about half an hour, the paste has a consistency like that of a chocolate mousse and its surface can be irreversibly deformed. If the experiment is conducted again with cement pastes made with a stiffening and a normal cement, it is when they are least shaken that they show the largest differences in rheological behaviours.

With a coaxial cylinders viscometer we have been able to reproduce and quantify the simple experiment described above. The more the paste is shaken during the measurement the more fluid it is. This phenomenon is called thixotropy and can be explained as follows: the cement paste is a flocculated suspension (especially in the case of stiffening), as a result of attractive forces between the cement particles; there is hence an internal solid structure inside the suspension and that structure can be more or less destroyed by agitation; such a structure progressively reappears when the paste is at rest.

4 Determination of an appropriate procedure to study cement pastes with rotational viscometers

Let us consider the shearing conditions of the cement paste at the time of concrete placing. When concrete is poured into a form, (and also during the course of the empirical tests which simulate the placing), there is some flow in the mass of concrete, but two sand grains which are close to each other at the beginning of the operation will probably still be close at the end. This means that the cement paste (which is located between the sand grains in concrete) is submitted to very small shearing during the placing of concrete. Let us note that the cement paste is submitted to much higher shearing during the mixing of concrete.

In order to simulate the shearing conditions of the cement paste during concrete placing, rheological measurements in viscometers must be carried at very slow speeds and during very short times.

Of course, such a conclusion would not hold for industrial problems related to hydraulic grouts or oil-well cement slurries.

When using a coaxial cylinders viscometer, a typical measurement is a cycle which consist of a regular increase of the shear rate from 0 to its maximum value and a regular decrease to 0 again. In the case of our viscometer (Contraves Rheomat 115), we can fix two parameters which are the maximum speed (number of revolution per minute) and the ramp (time in minutes which would be necessary to increase the speed from 0 to 780 rpm.). For a few values of these two parameters, we have calculated, in the table below, the maximum shear rate, the duration of the cycle and the total number of rotations during the cycle.

Max. Speed (rpm)	Max. shear rate (s-1)	ramp	Duration of cycle (s)	Number of rotations
780	1000	2	240	1560
350	450	2	107,7	314
112	144	2	34,5	32
22,4	28,8	40	138	25,8
22,4	28,8	2	6,9	1,29
11,2	14,4	2	3,45	0,32

The number of rotations of the inner cylinder gives a good idea of the degree of agitation of the paste during each cycle. From our experimental work, we consider that, in order to avoid a destruction of the structure of the cement pastes, the motion of the inner cylinder must be on the order of one complete revolution or less.

In the published studies carried on cement pastes with coaxial cylinders viscometers, different procedures have been used, but very often the maximum value of the shear rate is fairly high (400 to 1000 s^{-1}), and is most of the time larger than 100 s^{-1}. Such studies are then related to the rheology of cement pastes, the structure of which has been more or less destroyed.

In this work, we have chosen to use a cycle in which the inner cylinder rotates at a constant speed of 5 rpm during 5 seconds; hence, it turns by a total angle of 150° during the complete cycle. During the rotation the shear stress is continuously measured. Its value, which is nearly constant in the second half of the cycle, is very close to the true value of the yield stress of the cement paste. To study the rheology of cement pastes as a function of time such a cycle is carried out only every 15 minutes, the paste being left at rest in the viscometer between these cycles.

Practically, it is well-known that there is a continuous loss of workability of concrete with time during the dormant period. As it will be seen later on in this paper, with our procedure, we always observe a continuous thickening of the cement pastes as a function of time elapsed after mixing. We think that studies in which a decrease of the viscosity of the cement paste is measured at any time after mixing, are questionable as far as the extrapolation to concrete placing is concerned.

5 Rheological measurements with model suspensions of α alumina in water

In an earlier study, Chappuis (1982), we have used α alumina in water at different pH in order to study the rheology of inert model suspensions. Ultra-fine particles (between 0,1 and 1 μm) had been used in order to emphasize colloïdal phenomena. Depending on the value of the pH of the suspending liquid, such suspensions can be flocculated or dispersed and have quite different rheological behaviours.

In figure 1, we have reproduced the curve of the surface charge of α alumina as a function of pH from the work of Ballion and Jaffrezic-Renault (1982). The electric properties of the surface of α alumina particles in water have been investigated by many authors and a value of 8,5 is most often given for the point of zero charge. For pH values between 6,5 and 10,5 the absolute value of the surface charge of the particles is too small for the repulsive double layer interactions to overcome attractive van der Waals attraction forces and the suspensions are flocculated and have poor flowing properties.

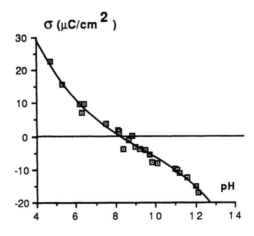

Figure 1: Surface charge of α alumina as a function of pH

For pH values smaller than 6,5 or higher than 10,5 the suspensions are dispersed and are rather fluid.

In the present study, a pure α alumina powder with a granulometry roughly in the same range as that of cements has been used: particles diameter was comprised between 1 and 50 μm with a median diameter of about 10 μm. Suspensions at different water to solid ratios have been prepared at two different pH values: 4,5 and 8,5. These suspensions have been studied in our viscometer with the same procedure as that used for the cement suspensions: measurement of the torque under conditions of very low shear (rotation of the inner cylinder at 5 rpm during 5 seconds, such a cycle being repeated every 15 minutes). In such conditions, and with all the tested suspensions the value of the shear

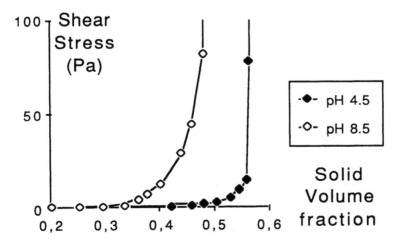

Figure 2: Rheological properties of α **alumina suspensions** as a function of solid volume fraction at two different pH

stress measured during the second cycle and the following ones is the same. Such values have been used to draw the curves the shear stress as a function of the solid volume fraction of the suspensions in the two pH conditions (see figure 2).

From the observation of figure 2, it can be seen that even with not purely colloïdal suspensions, the rheology is highly dependent on the state of flocculation of the suspension (dispersed at pH 4,5 and flocculated at pH 8,5). As the size of the particles of the model alumina suspensions used in this experiment is very close to that of cements, it can be inferred that interparticle forces and colloïdal phenomena have to be taken into account to improve our understanding of the rheology of cement pastes.

6 Rheological measurements on aluminous cement pastes during the dormant period

We have studied the rheology of aluminous cement pastes made with a stiffening cement and with a normal cement. The stiffening aluminous cement is the same as the one used in part 3 of this paper and the normal cement is a commercial aluminous cement (Ciment Fondu Lafarge). Different water to cement ratios (mass of water / mass of cement) have been used: 0,225 0,25 0,30 0,35 0,40 0,45 and 0,50 for the normal cement and 0,30 0,50 and 0,60 for the stiffening cement.

We have used the following procedure: cement pastes are prepared by mixing the cement with tap water for 3 minutes in a laboratory mixer at a medium speed (Kenwood mixer with a vertical paddle having an eccentric movement). Then the paste is poured into the viscometer and the first cycle (at 5 rpm during 5 seconds) starts 5 minutes after the first contact of cement and water. The paste is left at rest in the viscometer and an other cycle is conducted on the same sample every 15 minutes during 90 minutes.

During each experiment we have also recorded the temperature of the same cement paste stored in a small isothermal container. We have never noticed any temperature increase which would have been significant of the beginning of the hydration reactions:

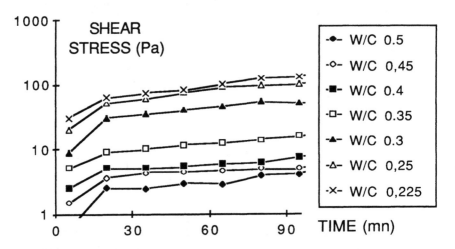

Figure3: Rheological behaviour during the dormant period of cement pastes made with a **normal aluminous cement**; each curve is related to a given water to cement ratio; the shear stress scale is logarithmic

8

consequently all our measurements on aluminous cement pastes have been conducted during the dormant period.

The results are presented on figure 3 for the normal cement pastes and on figure 4 for the stiffening cement pastes; on these figures, the scale of the shear stress is logarithmic.

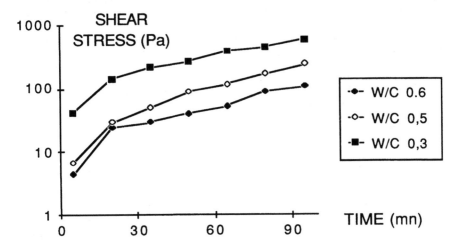

Figure 4: Rheological behaviour during the dormant period of cement pastes made with a **stiffening aluminous cement**; each curve is related to a given water to cement ratio; the shear stress scale is logarithmic

From these figures we can draw the following conclusions:
- during the dormant period all pastes exhibit a continuous thickening with time.
- the experimental points seem to be on parallel straight lines. The slope of these lines which is a measure of the relative thickening of the cement pastes with time is higher with the stiffening cement. For a given cement the relative thickening of the cement pastes is independent of the water to cement ratio.
- the first point of each set of data does not fit with the experimental straight lines. On each curve, the first point represents the measured value of the shear stress only 2 minutes after the end of the mixing. During these 2 minutes the paste has been introduced into the annular space of the viscometer. Let us recall that the cement pastes are thixotropic; consequently their internal stucture is damaged at the time of the first cycle and as a result the measured shear stress is smaller than that expected from the straight lines. As the second point of each set of results seems to be on the straight lines, we can deduce that the cement pastes have completely recovered their internal structure after 15 minutes at rest.
- by comparing the values of the shear stresses for the two cements, we can notice that the normal cement pastes are one order of magnitude more fluid than the stiffening cement pastes. For instance, at the same water to cement ratio (0,3) the shear stesses are ten times as large with the stiffening cement. We can also notice that a given shear stress level (for instance 100 Pascals at 90 minutes) is obtained with a water to cement ratio of 0,25 for the normal cement paste and with a water to cement ratio of 0,6 with the stiffening cement.

7 Rheological measurements on Portland cement pastes at different W/C ratios

We have used a commercial blended Portland cement CPJ 45 from our plant Lafarge in Le Teil. We have studied the rheology of Portland cement pastes at different water to cement ratios (mass of water / mass of cement): 0,275 0,3 0,4 and 0,5. A test has also been conducted at W/C = 0,6 but the results are not reported here because there has been some significative bleeding during the test.

We have used the usual procedure: cement pastes are prepared by mixing the cement with tap in our laboratory mixer (30 seconds at a low speed and then 2 minutes and 30 seconds at a medium speed). Then the paste is poured into the viscometer and the first cycle (at 5 rpm during 5 seconds) starts 5 minutes after the first contact of cement and water. The paste is left at rest in the viscometer and an other cycle is conducted on the same sample every 15 minutes during 4 hours at least.

The results are presented on figures 5 and 6; on figure 5 the scale of the shear stress is linear and on figure 6 it is logarithmic.

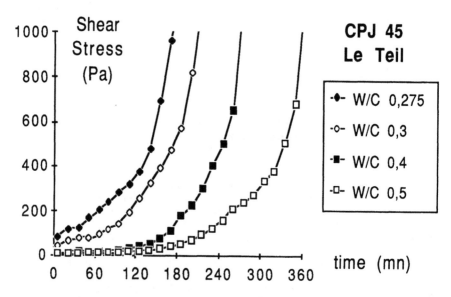

Figure 5: Rheological behaviour of **portland cement pastes;** each curve is related to a given water to cement ratio; the shear stress scale is linear

During the rheological tests, we have also recorded the temperature of a portion of the same cement pastes stored in a small isothermal container. These temperature recordings have been reported on figure 7.

From these figures, it can been seen that the transition between the dormant period and the thickening related to the hydration reactions is very progressive. The first increase in temperature and shear stress related to these reactions can be detected at about 120 mn for the W/C = 0,5 paste, at about 90 mn for the W/C = 0,4 paste, at about 60 mn for the W/C = 0,3 paste and at about 45 mn for the W/C = 0,275 paste. After this transition the increase in consistancy of the cement pastes is exponential as shown by the straight line appearance of the curves of figure 6.

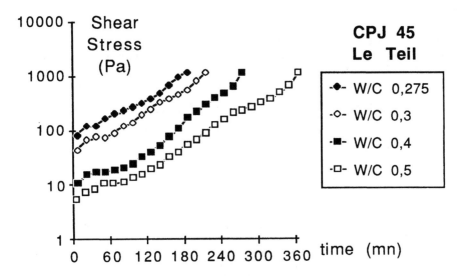

Figure 6: Rheological behaviour of **portland cement pastes;** each curve is related to a given water to cement ratio; the shear stress scale is logarithmic

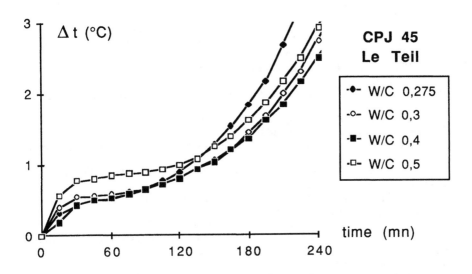

Figure 7: Increase in temperature of **portland cement pastes** when stored in a small isothermal container; each curve is related to a given water to cement ratio

8 Conclusion

The rheology of cement pastes and concrete is highly related to the state of flocculation of the cement particles. Although cement pastes are not pure colloïdal systems, we think that interparticle forces play a very important role as summarized in the following table.

FORCES BETWEEN PARTICLES	ATTRACTIVE	REPULSIVE
STATE OF THE SUSPENSIONS	FLOCCULATED	DISPERSED
RHEOLOGY	THICK	FLUID

Under agitation a cement paste is fluidified as a result of the destruction of its internal structure related to its more or less flocculated state.
During concrete placing, the cement paste which is located between the sand grains, is submitted to very small shearing. Then, we think it preferable to use very low shearing conditions to study cement pastes in viscometers.

REFERENCES

Adamson, A.W. (1982) **Physical Chemistry of Surfaces** 4th ed, J. Wiley N.Y.
Atzeni, C., Massidda, L. and Sanna, U.(1983) **Il Cemento** 80, 141.
Ballion, D. and Jaffrezic-Renault, N. (1982) **Proceedings 3° Meeting Société Générale d'électrochimie** Lyon.
Banfill, P.F.G. and Saunders, D.C. (1981) **Cem. Concr. Res.** 11, 363.
Banfill, P.G.F. and Gill S.M. (1986) **8th Intern. Congress on the Chemistry of cement,** Rio de Janeiro, vol.6 p. 223 .
Baron, J (1982) in **Le béton Hydraulique** p. 131 Presses Ecole Nationale des Ponts et Chaussées, Paris.
Chappuis, J. (1982) in **Concrete Rheology** p. 38 ed. J. Skalny, Materials Research Society.
Chappuis, J.(1986) in **Proceedings of the 8th Int. Congress on the Chemistry of Cement,** Rio de Janeiro, vol 6 p. 544.
Chappuis, J. (1988) in **Proceedings of the Xth International Congress on Rheology,** Sydney, vol. 1 p. 247.
Chappuis, J. Bayoux, J.P. and Capmas, A. (1989) in **UNITECR '89 Proceedings** Anaheim Ca. vol 2 p.1171-1182.
Costa, U. and Masazza, F. (1986) in **Proceedings of the 8th International Congress on the Chemistry of cement,** Rio de Janeiro, vol. 6 p. 248.
Derjaguin, B. and Landau, L. (1941) **Acta Physicochim.** 14, 633.
Furlong, D.N., Yate, D.E. and Healy, T.W. (1981) in **Electrodes of Conductive Metallic Oxides** Part. B., Trasatti S. Ed. Elsevier.
Hamaker, H.C. (1937) **Physica** 4, 1058.
Lapasin, R. Papo, A. and Rajgelj, S. (1983) **Rheol. Acta** 22, 416.
Tatterstall, G.H. and Bloomer, S.J. (1979) **Magazine of concrete research** 31, 202.
Roy, D.M. and Asaga, K. (1979) **Cem. Concr. Res.** 9, 731.
Roy, D.M. and Asaga, K. (1980) **Cem. Concr. Res.** 10, 387.
Verwey, E. and Overbeek, J. Th. G (1948) **Theory of Stability of Lyophobic Colloids,** Elsevier Amsterdam.

EFFECT OF FLY ASH ON THE RHEOLOGICAL PROPERTIES OF CEMENT PASTE

F. SYBERTZ and P. REICK
Institute for Building Research, Aachen University of
Technology, Aachen, Federal Republic of Germany

Abstract
The influence of fly ash on the rheological properties of cement
pastes was investigated using a Couette rheometer. The yield stress
τ and the initial viscosity η_0 were determined from the load branch
of the flow curve as decisive parameters. Tests were carried out with
two different fly ashes, varying the solids concentration by volume C_v
and the fly ash/solids ratio by volume ϕ. The measured value curves
were approximated in the form of an equation.
Keywords: Fly ash, Cement paste, Rheology, Concentration of solids

1 Introduction

Fly ash has long been employed for concrete production. One reason for
its widespread use is the resulting improvement in the properties of
fresh concrete. The effect on concrete workability has been investigated by a number of authors. It is well known that the fineness and
loss on ignition of fly ashes primarily affect the water requirement.
Nevertheless, the influence of fly ash on fresh concrete is still not
fully understood.

The influence of the cement and of additives and admixtures on rheological properties can be investigated to advantage in cement pastes,
where the principal effects are considerably more pronounced than in
concrete. It should, however, be noted that not all influencing factors act in the same way, primarily due to the different size distribution of particles in concrete and cement paste.

Although many rheological tests have in the past been carried out
on cement paste, few confirmed results are available (Tattersall and
Banfill 1983). This is mainly due to the fact that cement suspensions
show a very complex rheological behaviour and test conditions can com-
plicate the measurements. For example, a series of constraints must be
observed when testing cement paste, if useful results are to be obtained (vom Berg 1982, Tattersall and Banfill 1983).

It is known from numerous investigations that both the yield stress
and the initial viscosity are exponential functions of the water/ce-
ment ratio w/c (Tattersall and Banfill 1983). The rheological properties of suspensions are, however, characterized not by the mass proportion but by the volume proportion of solids. It is for this reason

13

that vom Berg (1979, 1982) chose the solids concentration by volume:

$$C_V = \frac{V_c}{V_w + V_c} \qquad (1)$$

where V_c = the volume of the cement and V_w = the volume of water in preference to the water/cement ratio as a parameter in his study. The influence of C_V on yield stress was approximated in an equation as

$$\tau_o = P1 \cdot e^{P2 \cdot C_V} \qquad (2)$$

The same relationship applies to initial viscosity, with different values for the parameters P1 and P2.

The fineness and C_3A content of the clinker were investigated as influencing factors on the cement at the Institute for Building Research (vom Berg 1979, 1982, Backes and vom Berg 1984). As the specific surface and C_3A content increased, there was a rise in τ_o and η_o. The influence of the C_3A content of the cement is dependent less on total quantity than on the amount going into solution during the first minutes. Both parameters affect P1 in Equation (2), whereas P2 remains unchanged. Vom Berg (1979, 1982) determined exponent P2 at 26.0 for τ_o and 18.8 for η_o for cements with specific surfaces ranging from 130 to 690 m²/kg. The cements had been produced from a basic cement by separation into particle fractions and subsequent mixing to various particle size distributions. Backes and vom Berg (1984) determined exponent P2 at 20.5 for τ_o and 19.5 for η_o for 12 cements with C_3A contents of between 0 and 11.2 % by mass and specific surfaces ranging from 300 to 520 m²/kg. The cements were produced from 4 industrial clinkers by grinding in a laboratory mill.

Table 1. Chemical and physical properties of the cement
and the fly ashes

Item	cement PC 35	fly ash FA I	fly ash FA II
Chemical compositon (%)			
- SiO_2	19.9	46.2	51.2
- Al_2O_3	5.77	26.0	27.6
- Fe_2O_3	2.63	11.8	8.56
- CaO	64.2	4.71	2.07
- MgO	1.36	2.92	2.21
- K_2O	0.84	3.09	3.93
- Na_2O	0.21	0.48	0.43
- SO_3	2.93	1.36	0.58
- Loss on Ignition	2.66	1.36	3.33
Fineness			
- Density (kg/m³)	3100	2320	2380
- Surface area, Blaine (m²/kg)	288	309	332

2 Experimental

Materials and Mixtures

The tests were carried out using an industrially-produced class 35 portland cement (PC) and two bituminous coal fly ashes from different power plants. Table 1 compares some of the chemical and physical properties.

Fly ash FA I came from a dry-bottom furnace operating at furnace temperatures of between 1100 and 1300 °C, fly ash FA II from a slag-tap furnace with furnace temperatures ranging from about 1600 to 1700 °C. Due to the higher temperature, fly ash FA II is composed of optimally-shaped spherical particles. By contrast, fly ash FA I contains some irregularly-shaped particles. The particle size distribution and SEM images of the fly ashes are available from a investigation by Schießl and Härdtl (1989), using the same basic materials. Particles larger than 125 μm were separated from the ashes and cement in order to ensure clear flow conditions at the rheometer gap width of 1 mm.

Fig. 1 provides a general overview of the test series which were carried out. The effect of substitution by volume was tested at 3 different concentrations of solids C_V of 0.392, 0.432 and 0.480. The fly ash/solids ratio by volume ϕ ranged from 0 to 100 %. Experiments with a water/(cement + fly ash) ratio w/(c+f) of 0.35 were also performed in order to determine the effect of substitution by mass. Fig. 1 clearly reveals the rise in C_V with increasing ϕ using substitution by mass.

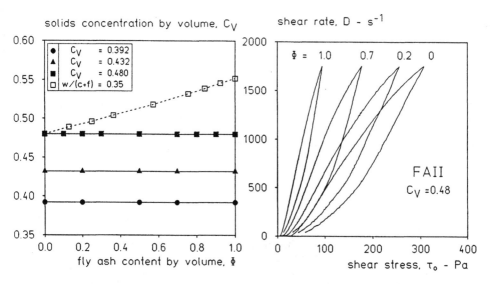

Fig. 1. Varied fly ash contents and solid concentrations in the test series

Fig. 2. Flow curves of cement paste at different fly ash contents

15

Equipment and Techniques

Flow curves of pastes were measured using a specially-built Couette rheometer. The measuring gap width was 1 mm and the ratio of the radii R_0/R_i = 1.027. To exclude sliding at the walls and the occurrence of plug flow, the measuring cylinder walls were serrated. Torque was measured using an stiff measuring array, enabling initial events to be measured with good accuracy and a wide shear stress range to be covered. Details of the rheometer and the testing procedure are given by vom Berg (1979, 1982).

Cement, fly ash and de-ionized water were mixed as follows, using a mixer complying with European Standard EN 196 Part 1: 1 min slow speed, 1 min high speed, 3 min waiting time and 1 min high speed. The shear test commenced 10 min after the beginning of the mixing process. In order to eliminate influences caused by the preliminary treatment, and in order to get equilibrium conditions for the flow curve, the paste was pre-sheared for 30 s at a low shear rate of 50 s^{-1}. The shear rate was then raised with a constant acceleration of about 29 s^{-2} to 1800 s^{-1}, and subsequently reduced to D = 0.

The flow curves shown in Fig. 2 resulted. The curves exhibit the form already observed by vom Berg (1979, 1982). Shear thinning is observable in the lower and slight shear thickening in the upper range. Vom Berg (1982) showed that the changed behaviour in the upper flow curve range at shear rates in excess of about 1000 s^{-1} is attributable to turbulence phenomena resulting from the extremely high shear rates.

The flow curves were evaluated by determining the yield stress τ and initial viscosity η_o = A/B. The lower flow curve range between $\dot{B}$ = 50 and 380 s-1 was approximated using the equation already employed in earlier studies (vom Berg 1979, 1982):

$$D = B \cdot \sinh \frac{\tau - \tau_o}{A} \qquad (3)$$

Other experiments

The spread at flow table according to German Standard DIN 1060 Part 3 was also determined on mortars. The reference mortar was mixed from 450 g cement, 225 g water and 1350 g standard sand, corresponding to a w/(c+f) ratio of 0.50 and a C_v of 0.392. The spread at flow table was 160 mm. In the fly ash mortars, the cement was substituted both by the same mass and by the same volume of fly ash.

3 Results

The results of the mortar tests are shown in Fig. 3. Substitution of the fly ashes FA I and FA II either by mass or by volume resulted in liquefaction. As expected, the effect of fly ash is greater with dosing by volume. The spread at flow table increases in roughly linear proportion to the fly ash content ϕ. Fly ash FA II has a significantly greater liquefying effect than FA I. As indicated in the tests carried out by Schießl and Härdtl (1989), this is related not only to the higher fine particle component but also to the more favourable particle shape of FA II.

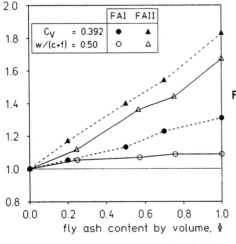

relative flow spread of mortar

Fig. 3. Effect of fly ash content on relative flow spread of mortar

fly ash content by volume, Φ

Figures 4 and 5 show the influence of solids concentration on rheological parameters for cement pastes with 100 % fly ash FA I or FA II and portland cement (PC) respectively. If yield stress or initial viscosity are plotted logarithmically, the values are approximately linear for both fly ashes, indicating that Equation (2) is applicable to fly ashes as well as cement.

Both the yield stress and the initial viscosity are distinctly lower for fly ash FA II than for the portland cement, whereas the flow behaviour of fly ash FA I differs much less from that of cement; the yield stress is somewhat lower and the initial viscosity somewhat

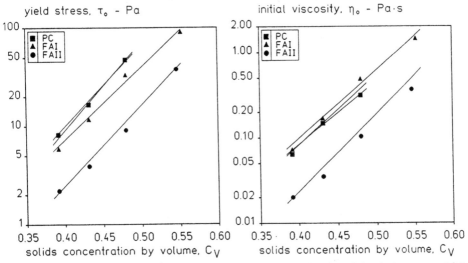

yield stress, τ_0 - Pa

initial viscosity, η_0 - Pa·s

solids concentration by volume, C_V

solids concentration by volume, C_V

Fig. 4. Relationship between solids concentration and yield stress

Fig. 5. Relationship between solids concentration and initial viscosity

higher than for PC.

The influence of the fly ash content by volume ϕ on τ_o and η_o is shown in Figures 6 to 9. It will be evident that the rheological parameters increase in linear proportion to the fly ash content.

An approximation in the form:

$$\tau_o = P1 \cdot e^{P2 \cdot C_V} \cdot (1- \phi) + P3 \cdot e^{P4 \cdot C_V} \cdot \phi \qquad (4)$$

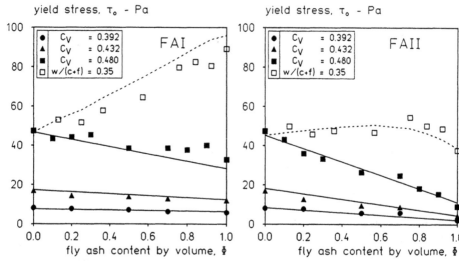

Fig. 6. Effect of fly ash content on yield stress, FA I

Fig. 7 Effect of fly ash content on yield stress, FA II

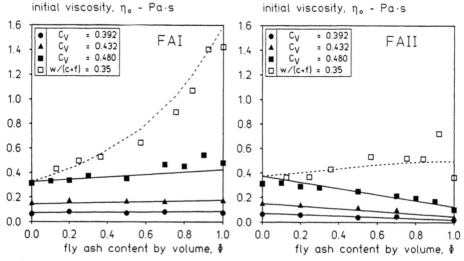

Fig. 8. Effect of fly ash content on initial viscosity, FA I

Fig. 9. Effect of fly ash content on initial viscosity, FA II

was therefore selected for τ_o or η_o. The equation consists of two
terms describing the properties of the cement and the fly ash respectively. At 0 % fly ash (ϕ = 0), only parameters P1 and P2 are relevant, for 100 % fly ash only parameters P3 and P4. Compensation calcu-
lations ln τ_o = f(C_V, ϕ) or ln η_o = f(C_V, ϕ) were carried out according
to the least squares method for the available data. The parameters de-
termined in each case and the corresponding coefficient of determination B are indicated next to Figures 10 to 13. The curves resulting
from Equation (4) are also shown in Figures 4 to 13. It will be apparent that the equation chosen is a good approximation of the measured-

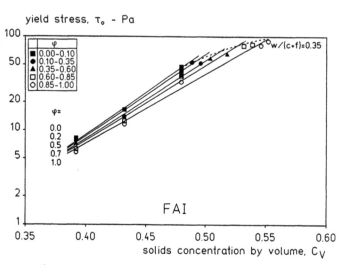

$$P1 = 2.31 \cdot 10^{-3}$$
$$P2 = 20.7$$
$$P3 = 7.96 \cdot 10^{-3}$$
$$P4 = 17.0$$
$$B = 99.89 \text{ %}$$

Fig. 10. Yield stress as function of C_V and ϕ

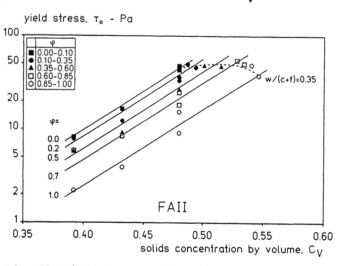

$$P1 = 4.75 \cdot 10^{-3}$$
$$P2 = 19.1$$
$$P3 = 1.29 \cdot 10^{-3}$$
$$P4 = 18.9$$
$$B = 99.80 \text{ %}$$

Fig. 11. Yield stress as function of C_V and ϕ

value curves. The difference between tests with a constant w/(c+f) ratio (broken line) and tests with a constant C_V is extremely clear.

It should be mentioned that compensation calculations for the two fly ashes FA I and FA II yielded different values for P1 and P2. This is related to the fact that only 3 measured values were available for pastes with 100 % portland cement. The number of measured values and the range of solids concentrations investigated are too low to permit more accurate determination of the parameters. It can, however, be inferred from Figures 4 and 5 that despite the differing parameters in the investigated range there are no large deviations between the two curves.

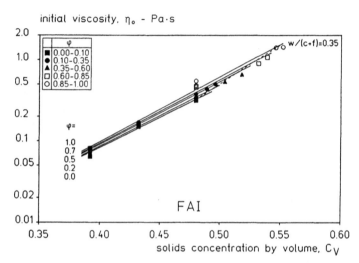

P1 = 8.97·10⁻⁵
P2 = 17.1
P3 = 6.25·10⁻⁵
P4 = 18.4
B = 99.56 %

Fig. 12. Initial viscosity as function of C_V and ϕ

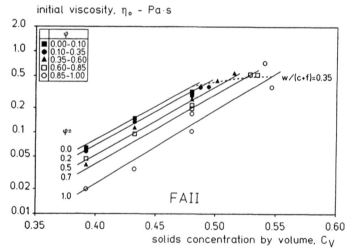

P1 = 4.04·10⁻⁵
P2 = 19.0
P3 = 0.54·10⁻⁵
P4 = 20.9
B = 99.39 %

Fig. 13. Initial viscosity as function of C_V and ϕ

4 Discussion

It is surprising that the linear correlation between ϕ and τ_0 or between ϕ and η_0 extends to high fly ash contents. Flatten (1973) observed an abrupt drop in the yield stress to zero at the transition to 100 % quartz powder in his tests with portland cement and ultra-fine quartz powder. Flow behaviour altered markedly when he employed the supernatant liquid of cement paste instead of distilled water to produce the pastes. Yield stresses for quartz powder were then higher than those for cement.

The flow behaviour of suspensions is closely related to the physical forces between the individual particles. Owing to the effects of surface forces, particles occur not individually but as agglomerates; they form a flocculate structure. A minimum shear stress τ_0 is required to destroy this network. The surface forces can be influenced in a variety of ways, for example through the destruction of the flocculate structure by addition of superplasticizer, causing the yield stress to fall to zero (Sybertz 1985).

Sedimentation tests were carried out with suspensions of the two fly ashes and of the cement at a water/solids ratio of 2.0 in order to demonstrate the existence of a flocculate structure. All sedimentation tests revealed a sharp division between the solid/liquid suspensions and the virtually pure liquid above them. This indicates that the particles do not sediment individually according to Stokes' law, but occur as loosely bound agglomerates produced by the interparticle forces, and fall to the bottom as a single mass.

The surface forces are strongly influenced by the type and concentration of the dissolved ions. Electrical conductivity and pH value were therefore measured on completion of the sedimentation tests (90 min). Results for PC, FA I and FA II showed conductivities of 16.3, 7.0 and 2.1 mS/cm and pH values of 12.5, 12.3 and 9.8 respectively. It is clear from these results that large quantities of ions go into solution even in pure fly ash mixtures. The conductivity and pH value of fly ash FA I are distinctly higher than for FA II indicating that more salts are dissolved.

These sedimentation tests show that in the case of pure fly ash sufficient electrolytes go into solution to create a flocculate structure. With pure quartz powder, on the other hand, these electrolytes have to be added with the mixing water. Owing to the formation of a flocculate structure, the relationship between the rheological parameters and the solids concentration is fundamentally the same for fly ashes as for cements. Equation (2), developed by vom Berg for pure cement pastes, can therefore also be applied to pure fly ash paste.

Figures 4 and 5 show that the curve gradients and thus parameter P2 in Equation (2) are not significantly different for fly ash and cement pastes. It is not yet possible to judge the extent to which this phenomenon may be coincidental. It does, however, appear that the differences in material properties manifest themselves specifically in factor P1 in Equation (2) in fly ash as in cement.

5 Conclusions

The following conclusions may be drawn from the test results:

(a) The flow behaviour of pure fly ash paste does not differ fundamentally from that of cement paste, provided that a flocculate structure can form. The necessary conditions are ensured in the range of solids concentrations investigated by the content of water-soluble salts in fly ashes.

(b) The approximation developed by vom Berg to describe the influence of the solids concentration by volume on the yield stress or initial viscosity (Equation 2) in the case of portland cement can also be applied to fly ashes. The influence of the material properties manifests itself principally in parameter P1.

(c) Both in mortar and in paste, the changes in the rheological parameters caused by substitution of the same volume of fly ash for cement are linear, i.e. the effects of fly ash and cement are additively superimposed.

(d) Fly ash FA II has a much stronger liquefying effect than FA I. Differences in fly ash properties can be estimated particularly clearly in 100 % fly ash pastes or mortars with no added cement.

References

Backes, H.P. and vom Berg, W. (1984) Rheological experiments on influence of chemical composition of cement upon flow behaviour of cement pastes, in **Proc. 9th Int. Congr. on Rheology**, Acapulco, Mexico, Vol. 2

vom Berg, W. (1979) Influence of specific surface and concentration of solids upon the flow behaviour of cement pastes. **Mag. Concr. Res.**, 31, 211-216

vom Berg, W. (1982) Zum Fließverhalten von Zementsuspensionen, **PhD thesis**, Aachen University of Technology

Flatten, H. (1973) Untersuchungen über das Fließverhalten von Zementleim, **PhD thesis**, Aachen University of Technology

Schießl, P. and Härdtl, R. (1989) The change of mortar properties as result of fly ash processing, in **Fly ash, Silica fume, Slag and Natural pozzolans in Concrete**, supplementary papers (ed. V.M. Malhotra), Trondheim, pp. 277-294

Sybertz, F. (1985) Einfluß von Fließmitteln auf die rheologischen Eigenschaften von Portlandzementsuspensionen. in **Baustoffe 85**, Bauverlag, Wiesbaden, Berlin, pp. 218-225

Tattersall, G. H. and Banfill, P.F.G. (1983) **The Rheology of Fresh Concrete**. Pitman Books Limited, London

3 THE EFFECT OF CONDENSED SILICA FUME ON THE RHEOLOGICAL BEHAVIOUR OF CEMENT PASTES

Y.P. IVANOV and T.T. ROSHAVELOV
Bulgarian Academy of Sciences, Sofia, Bulgaria

Abstract
The complex influence of condensed silica fume, superplasticizer and the mineral composition of cement on the rheological behaviour of fresh cement pastes has been studied, using mathematical theory of experiments. By the help of regression analysis for five factors, an optimizational equations has been established. The most significant role in reology of cement turned out to play the presence of condensed silica fume, used as an additive to the cement. A possible explanation of the anomaly of the viscosity of cement pastes, modified with this additive, has been proposed.

1 Introduction

In resent years, some attention have been given to the use of condensed silica fume (CSF) in concrete technology. The physical and chemical properties and the possible applications and limitations of this material in concrete has been well established (1,2,3,4). Moreover, some research have been made on the investigation of the workability of superplasticized microsilica-portland cement pastes (6) and concretes (5). However, the knowledge concerning the influence of CSF on the rheological properties of cement pastes and concretes is insufficient yet.
    The primary objective of the present study was to investigate the rheological properties of cement pastesmodified with CSF and superplasticizer using mathematical modeling. This approach would enable to estimate complex influence of CSF and other factors on fundamental rheological parameters of cement pastes.

2 Materials and methods

For preparation of the mixes studied two types of clinkers were emploied. One of them (Clinker-1: $C_3S-47,45\%$; $C_2S-25,92\%$; $C_3A-9,30\%$; $C_4AF-13,01\%$) was prepared with low content of C_3A and the other (Clinker-2: $C_3S-62,17\%$; $C_2S-14,93\%$; $C_3A-1,43\%$; $C_4AF-16,72\%$) with higher one.Different amounts of C_3A in the examined clinkers was reached by blending certain amounts of these two commercially available clinkers.Suitable amounts of gypsum (calculated in SO_3) was added to the clinkers. As an additive to the cement, CSF ($SiO_2-91\%$, B.E.T.$-16,3m^2/kg$) was used.The superplasticizer used was sulfonated melamine formaldehyde condensate, applied as solution with 29% wt solid concentration. The water/cement radio was varied in the range 0,35 to 0,55.

In agreement with the choosen plan of the experiment (saturated plan - CH_4), 22 mixes were examined. Prior to the experiments all materials were sifted through 0,2 mesh sieve. Pastes were prepared with different amounts of components (see Table 1) and mixed with a spatula for 90s. They were than placed in a mechanical blender and sheared at 300 rmp for 60s. Tests were started five minutes after water addition using rotational coaxial cylinder viscometer Rheotron-Brabender and rheograms were automaticaly registered with x-y-t Lauman recorder. All experiments were accomplished at $25\pm0,1$ °C. A complete hysteresis cycle was performed for 80s. The yield value was calculated from the initial reading of the scale and the value of viscosity was calculated at speed $145s^{-1}$, from the downcurve of the hysteresis loop. In each experimental point from the plan two tests, with one measurement each, were carried out.

Table 1.Variation limits of the components (factors)

Variation levels	X_1 w/c	X_2 SP,%	X_3 CSF,%	X_4 C_3A,%	X_5 SO_3, %
Mean quantity	0,45	1	7,5	5,37	2,5
Variation interval	0,02	0,2	1,5	0,79	0,2
Min. level	0,35	0	0	1,43	1,5
Max. level	0,55	2	15"	9;30	3,5

Data collected from the experiments were used in the regression analysis, applying the matrix of CH_4 plan (8). By means of adequate models (α=0,5), the interaction and the single influence of separate factors (components) were estimated. The regression coefficients were obtained by applying the least squares method.

3 Results and discussion

For the value of τ_o , the adequate polynomial model is:
τ_o = -118,5 - 72,3X_1 - 104,8X_2 +103,6X_3 - 46,1X_4 - 48,7X_5+ 75,4X_1X_2-77,4X_1X_3 + 38X_1X_4+ 42,4X_1X_5 - 107X_2X_3 +39,9X_2X_4 + 41,6X_2X_5 -54,5X_3X_4 -50,7X_3X_5 + 21,5X_1^2 + 19,3X_2^2 + 158,7X_3^2 + 38,9X_4^2 + 37,1X_5^2 ...(1)

For the value of η ap., the polynomial equation is:

ηap = - 300,8 - 357,6X_1 - 553,4X_2+ 575,4X_3 - 80,8X_4 - 22,4X_5 + 293, 7X_1X_2 - 329,4X_1X_3 + 14,7X_1X_4 + 57X_1X_5 - 528,4X_2X_3 + 48,9X_2X_4 + 21,2X_2X_5 - 151,1X_3X_4 - 29,3X_3X_5+ 175,2X_4X_5 + 114,4X_1^2 + 152,8X_2^2+664,6X_3^2 +159,2X_4^2 + 213X_5^2 ...(2)

Equations 1 and 2 allow to make detail interpretations and different optimisatial problems to be solved. From the models it is seen that the content of individual components in the paste affect the values of the rheological properties in different directions and depends on the level of the remaining factors. The verification of adequacy of the regressional coefficients shows that all of them are signi-

icant, which means that each factor (component) has substantial con-
ribution to flow behaviour and rheological properties of cement pastes
owever, quasimonofactoral m odels show, that separate factors can be
lassified, depending on their potent, as follows: $X_3 > X_2 > X_1 > X_4 > X_5$.
    As it was expected, the influence of water/cement ratio (X_1) and the
ontent of superplasticizer (X_2) is nearly linear and leads strongly to
ecrease of τ_0 and η_{ap} of the paste, when their levels (values) are
ncrease. The range of variations of the factors C_3A (X_4) and $SO_3(X_5)$,
.e. their influences, are significantly lower, than the above mentio-
ed. However, when the percentage content of these two factors increa-

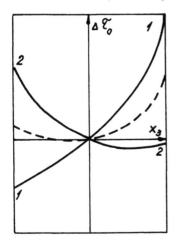 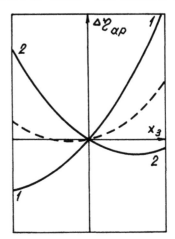

Fig.1. Quasimonofactoral models for the influence CSF (X_3) on:
a)   and   b)   of the paste:
       1 - upper limit of influence,
       2 - lower limit of influence.

se, observations show initial drop in the values of   and   followed
by increase of the rheological characteristics. Aforesaid is in agreement with the results, obtained in (7). This means, that even in case
of presence of CSF in cement, an optimum content of C_3A and SO_3 exists
from rheological point of view.
    Models (equ.1 and 2) show that the interaction of separate compone-
nts (factors) is strongest between the content of superplasticizer and
CSF in the paste.
    From technological point of view, the influence of CSF content on
the flow behaviour of the paste is of particular interest. Figure 1-
a) and b) show the relative influence of this component (X_3) on the value of τ_0 and η respectively, when the remaining one are optimized,i.e.
accept an extremal values.
    It was established, that additive of CSF (X_3) exerts most significa-
nt influence on rheological characteristics of cement paste. The degree
of its influence, as an absolute value exceeds the influence of the remaining facrors more than twice. When the CSF content in the material
increase to a certain value, τ_0 and η_{ap} decrease, followed by rigid increase of the rheological characteristics. The latter increase is essenti-

ally intensive in the range 7,5 to 15% CSF. The fore mentioned pheno-menon can be explained using the idea for anomaly of the viscosity (9) of suspensions with polydispersal solid phase. The observed minimum of the viscosity c an be ascribe to a state of suspension, in which the mo-st compact packing of the grains of cement and CSF is reached, due to their generaly different granulometry compositions and specific surfa-ces.

4 Conclusions

From the adduced above, it makes clear, that the separate components have to be selected after a complex estimation of the examined proper-ties or on the basis of dominant property, i.e. it is not possible the amount in percent of one component to be fixed previously, if the qua-ntity of the remaining ones are variable.

The most powerful factor (component) from the studied ones, with re-spect to rheology turned out to be the CSF used as an additive to the cement. Its presence rigidly changes the value of τ_0 and η_{ap} of the pas-tes, wich is most sensible in the range 7,5 to 15% CSF and above.

Even in case of presence of CSF in cement, an optimum content of C_3A and SO_3 exists from rheological point of view. Manifested anomaly of the viscosity, in this case, is connected with state of paste in which most compact packing of grains is reached.

5 References

1.Malhotra, V.M. and Carette, G.G., Concr.Intern.:Design and Constr., vol.5,No5, 1983.
2.Parker, D.G., Concrete, Oct., 1986.
3.Ollivier, J.P. at al., VIII Int. Congr. Chim. of Cem.,Sept.,1986.
4.Final report on Siliceous by - products for use in concrete, 73-SBC RILEM Committee, Mater. et Constr., Vol.21,No 121, 1988.
5.Mangialardi, T. and Paolini, A. E., Cem. Concr. Res.,Vol.18,351,1988.
6.Chiocchio, G.,Mangialardi,T. and Paolini,A.E.,ll Cementó (2),69,1986.
7.Ivanov,Y. and Stanoeva, Silicates Industr.,9,199,1979.
8.Voznesensky, V.A., Contemporary optimisational methods for composite materials, Kiev, Budivelnik, 1987.
9.Ivanov, Y., Microrheology of fluid silicate and polymer composites, D.Sc.Thesis, Bulgarian Academy of Sciences, 1988.

4 THE INFLUENCE OF CHEMICAL COMPOSITION OF CEMENT ON THE RHEOLOGICAL PROPERTIES

S. GRESZCZYK
Institute of Mineral Building Materials, Opole, Poland
L. KUCHARSKA
Technical University of Wroclaw, Wroclaw, Poland

Abstract
The research results of rheological properties of industrial clinker and cement differing in chemical and mineralogical composition have been discussed. Referring to the flow curves and shear stress changes in time the differences in rheological properties were determined. The consistency of cement pastes possessing similar mineralogical compositions and with different alkalis content, was found to be related to the 28-day strength. The C_3A content in these cement pastes is not decisive factor as regards the consistency of cement pastes; rather, these are alkalis built-in clinker phases which are dominant. Those cements of similar mineralogical composition the pastes of which are characterized by the lowest consistency during the initial stage of hydration, are found to reveal the highest strength.
Keywords: Clinker, Cement, Consistency, Alkalis, Strength.

1 Introduction

Numerous studies of the rheological properties of cement pastes have proved these properties to be up to many factors, the most frequently cited are the water/cement ratio (w/c), Collepardi (1971), specific surface area, mineral composition, Berg (1979), Odler et al. (1978), conditions during the measurements and their duration, Banfill (1981a). It has been shown that the most important are the w/c ratio and specific surface area. Studies performed on cement pastes of different chemical composition indicated this factor bears a less effect on the rheology than w/c ratio and/or fineness of cement, Odler et al. (1978), Banfill (1981b). The most meaningful influence of chemical composition is revealed in the retardation of the process of aluminates hydration caused by calcium sulphate, Bombled (1980). Due to the considerable reaction rates of these processes they are very important when determining rheological features of cement pastes at early hydration stages and also affect the development of rheological properties.

Examination of a great deal of important factors influencing the rheology of cement pastes and difficulties in the univocally determination of their contribution in the modeling of these properties indicates for the complexity of water-cement system. This also explains that a quantitative characterization of the effect of the factors mentioned above on the rheological bahaviour of cement suspensions is yet to be achieved.

This study comprises the results of rheological properties of both clinker and corresponding cement pastes of different chemical and mineral composition.

2 Experimental

Experiments were carried out using industrial clinkers and cements. All cements contained equal amounts of the setting control agent 5% wt of $CaSO_4 \cdot 2H_2O$. Both clinkers and cements were ground to the specific surface area of 300 ± 20 m^2/kg (Blaine). Table 1 presents the mineralogical composition of clinkers, contents of alkalies and free CaO, and the sulphatization degree of alkalis which is defined as the ratio of % content of SO_3 to the content of alkalies expressed as Na_2O , Richartz (1986).

Table 1. The mineralogical, chemical composition (% wt) and the degree of sulphatization DS of the clinkers(%)

Clinker	C_3S	C_2S	C_3A	C_4AF	Na_2O	K_2O	Na_2O_e	SO_3	CaO	DS
A_1	53.0	20.0	15.4	7.5	0.20	1.20	0.99	0.70	2.1	55
A_2	55.0	21.3	15.0	7.9	0.20	1.00	0.86	0.40	2.5	36
A_3	54.1	21.0	15.4	7.5	0.10	0.90	0.69	0.60	2.4	67
B_1	51.8	25.5	11.0	7.9	0.10	1.20	0.89	0.50	1.1	40
B_2	57.3	20.4	8.5	8.8	0.20	0.50	0.53	0.90	1.2	132
B_3	57.0	23.0	8.5	8.0	0.10	1.00	0.76	0.90	1.0	92
C_1	50.0	21.0	13.6	7.9	0.20	1.20	0.99	0.30	1.2	23
C_2	51.0	22.0	12.0	9.7	0.20	0.70	0.66	0.40	1.2	47
C_3	48.9	25.7	12.8	9.4	0.10	0.60	0.50	0.20	0.9	32
D_1	64.7	8.5	14.9	7.9	0.20	0.70	0.66	0.50	2.6	59
D_2	62.2	10.1	14.0	8.5	0.20	0.70	0.66	0.40	2.1	47
E_1	57.2	20.1	7.7	8.5	0.10	1.10	0.82	0.70	0.7	75
E_2	50.0	26.0	8.6	9.3	0.20	1.30	1.06	1.50	0.8	124
F_1	58.6	14.9	9.3	13.4	0.10	0.70	0.56	0.70	1.0	97
F_2	56.0	17.0	10.2	12.5	0.20	0.70	0.66	0.60	0.9	70
F_3	60.2	14.9	8.6	12.9	0.10	0.50	0.43	0.30	0.8	61
G_1	57.0	21.0	8.9	7.9	0.30	0.50	0.63	0.40	0.6	49
G_2	61.5	19.5	10.3	7.0	0.10	0.20	0.23	0.10	0.6	33
H	53.9	21.7	10.5	10.7	0.10	0.50	0.43	0.30	0.8	61
I	62.0	13.5	7.5	12.5	0.10	0.40	0.36	0.20	2.7	47
J	36.0	37.0	4.7	16.7	0.10	0.85	0.66	1.00	0.8	117

Measurements were taken by the use of a rotary viscosimeter Rheotest RV-2. In order to reduce sliding and not
uniform shear in the measuring slot as well as to decrease
the sedimentation of cement particles, the geometry of the
both coaxial cylinders was modified. Measurements were performed at a constant temperature of 21°C, the w/c was 0.50
and 0.40 for clinker and cement pastes respectively, and
was also maintained constant. Preparation of the test pastes included manual mixing (for 3 minutes) immediately
after the clinker and/or cement contacted with water, next
the suspension was fed into the viscosimeter and after 8
minutes of contact with water a 1 minute mixing was performed at a shear rate 146 s^{-1} after which the suspension was
left for 1 minute to rest before the start of measurements.
Thus, measurements were started after 10 minutes of the
contact of cement with water.

The rheological behaviour of the tested clinker and
cement pastes was determined basing on the obtained flow
curves and results on stress changes measurements taken
within 1 hour period at a constant shear rate ($\gamma = 46$ s^{-1}).
The flow character was determined from the run of curves
obtained for stress measurements when the shear rate was
increasing and decreasing within the range of 0 to 146 s^{-1}.

3 Results and discussion

In the tested clinkers (table 1) the C_3A content varied
from 4.7 to 15.4 % wt. Alkali content was 0.20 to 1.30 % wt
for K_2O and 0.10 % to 0.30 % wt for Na_2O. The sulphatization
degree of alkalies in clinkers ranged from 23 to 132 %.
From the analysis of the sulphatization degree figures it
follows that in clinkers B_2, E_2, J the alkalis can be present only as sulphates due to the SO_3 content. In other
clinkers a considerable amount of alkalies is incorporated
into the clinker minerals.

Clinker pastes containing the largest quantities both of
alkalies built-in into the clinker phases (low sulphatization degree of alkalies) and C_3A revealed in the rheological tests the highest consistency and its increment in time
(Table 2, Fig. 1, 2, 3 and 4). The raise of clinker paste
consistency due to the increase of alkali content in clinker is more distinctive than due to higher contents C_3A .
The correlation coefficient (r) between τ_{30} and Na_2O_e (i.e.
content of alkalies in clinker expressed as Na_2O) is $^e 0.75$,
between τ_{60} and Na_2O_e is 0.79, whereas the correlation
coefficient between τ_{30} and the C_3A content is 0.43 and
between τ_{60} and C_3A contend is 0.41. In the statistic analysis the four clinkers (B_2, E_1, E_2, J) containing alkali quantities from 0.5 to 1.0 % wt Na_2O_e and of high sulphatization
degree (~ >100%). Pastes obtained from these clinkers exhibited a considerable increase in consistency during the

initial 10 minutes of hydration, thus making the rheological measurements impossible to be carried out.

Table 2. Dynamic yield values τ_B, plastic viscosity n_{pl} and shear stress values τ_{30} and τ_{60} after 30 and 60 minutes for clinker and cement pastes, and 28-day strength values W_{28} for cement pastes

Clinker Cement	Clinker				Cement				
	τ_B	n_{pl}	τ_{30}	τ_{60}	τ_B	n_{pl}	τ_{30}	τ_{60}	W_{28}
	(Pa)	(mPas)	(Pa)	(Pa)	(Pa)	(mPas)	(Pa)	(Pa)	(MPa)
A_1	70	660	96	150	70	1350	67	103	30.7
A_2	63	520	111	130	52	650	42	92	32.0
A_3	59	410	81	115	42	370	34	60	42.6
B_1	48	500	90	140	53	910	53	90	28.0
B_2	-	-	-	-	36	450	35	60	35.4
B_3	18	300	74	110	38	500	50	70	32.4
C_1	50	660	100	145	-	-	-	-	28.0
C_2	23	350	70	116	47	450	62	130	29.3
C_3	38	600	80	115	32	360	46	90	44.3
D_1	60	320	76	117	20	360	44	75	32.8
D_2	26	270	24	56	110	1100	95	142	28.0
E_1	-	-	-	.-	23	270	35	60	28.3
E_2	-	-	-	-	40	420	35	65	30.2
F_1	61	410	65	95	35	360	30	70	33.5
F_2	40	430	80	120	65	550	63	111	35.1
F_3	22	380	68	100	32	250	26	67	38.0
G_1	25	440	60	113	35	390	25	60	40.5
G_2	9	320	24	68	55	500	30	50	46.5
H	12	340	60	90	49	430	50	80	28.7
I	10	200	45	80	40	380	33	70	34.8
J	-	-	-	-	25	340	20	37	29.0

Relations found between consistency of clinker pastes and the contents of alkalies and C_3A do not occure in the case of cement pastes. A comparison of rheological data on clinker and cement pastes of approximate equal C_3A content (taken from the same cement plant) has revealed the effect of the quantity and type of alkalies on the rheological properties of fresh cement pastes. Including the sulphatization degree of alkalies allows to draw more detailed conclusions as far as the influence of the type of alkali compounds upon the reactivity and rheology of clinker is concerned.

From the position of the flow curves obtained for three clinker pastes A_1, A_2, A_3 (prepared from clinkers taken from the same cement plant) and from determined dynamic yield values and plastic viscosities it follows (table 2) that these clinkers belong to the most reactive ones.

Moreover, the fact that clinker pastes A_1 and A_2 reveal higher consistency than A_3 allows to assume they are more reactive compared to the latter one. Taking into account the alkali content and sulphatization degree in the studied clinkers A_1, A_2 and A_3 one can find out that higher reactivity is produced by clinkers containing more alkalies, i.e. A_1 and A_2 (0.99 and 0.86% wt of Na_2O_e, respectively) and of a lower sulphatization degree (55 and 36%, respectively).

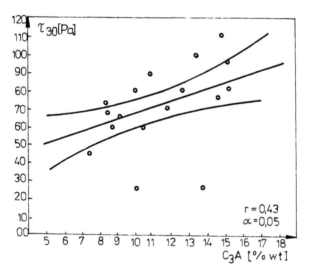

Fig. 1. Shear stress dependence upon C_3A content for clinker pastes after 30 minutes hydration.

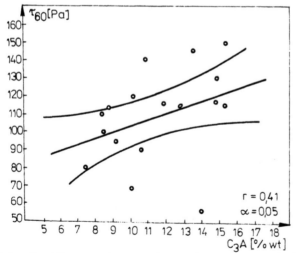

Fig. 2. Shear atress dependence upon C_3A content for clinker pastes after 60 minutes hydration.

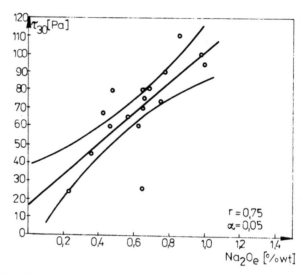

Fig. 3. Shear stress dependence upon alkali content
for clinker pastes after 30 minutes hydration.

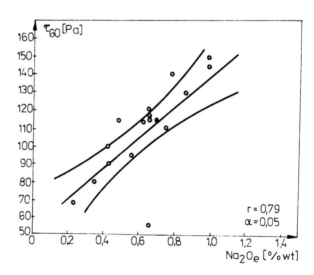

Fig. 4. Shear stress dependence upon alkali content
for clinker pastes after 60 minutes hydration.

Cement pastes A_1 and A_2 which contain larger amounts of
built-in alkalies are characterized by a high consistency.
This indicates for a low efficient control over the hydration process of C_3A by gypsum in the case of the most
reactive clinkers. However, the same gypsum addition produces the highest flowabilities of the least reactive clinker
A_3.

Lower content of built-in alkalies (A_3) corresponds to the lower consistency and its slight changes in time (Fig. 5). Formation of fine crystalline ettringite found in the cement paste A_3 indicates for a proper control of the hydration process by gypsum what is confirmed by the low paste consistency. The occurence of needle-like ettringite crystals and monosulphate in the cement paste of a high content of built-in alkalies (A_1 and A_2) indicates the gypsum to be used up faster and explains the reason of rapid consistency increase (Grzeszczyk and Kucharska, 1990). Results of hydration heat evolution have confirmed the lack of efficient control over the hydration process by gypsum for a cement containing larger quantities of alkalies incorporated into the clinker phases (Grzeszczyk, 1989), what has been found in the rheological tests.

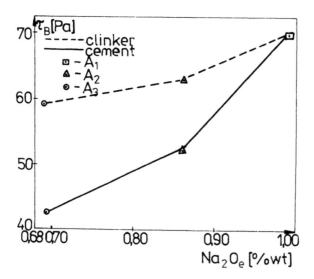

Fig. 5. Relation between dynamic yield values and alkaline contents for clinker and cement pastes A_1, A_2 and A_3.

A similar behaviour towards water as in the case of clinkers and cements A_1, A_2 and A_3 have also revealed clinkers and cements B_1, B_2 and B_3 and C_1, C_2, C_3. The highest reactivity exhibit clinkers which contain the largest amount of alkalies (B_1, C_1) and have got the lowest sulphatization degree. Cement pastes prepared from these clinkers have shown a high consistency like cements A_1 and A_2. In the case of cement paste C_1 which has had the highest alkali content and the lowest sulphatization degree (23%) from among the tested clinkers there is a substantial increase of consistency, thus making the rheological test impossible to perform already after 10 minutes of hydration.

This behaviour also evidences a low efficiency of gypsum as a hydration control agent in this paste.

28-day compressive strengths of the tested cement pastes A, B and C have shown a similar trend of changes. The strengths are the greater the lower is the paste consistency at the initial hydration stage (Fig. 6).

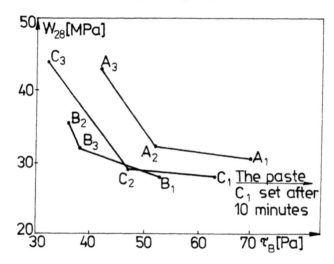

Fig. 6. Relation between 28-day compressive strengths and dynamic yield values for clinker and cement pastes.

Analysis of the flow curves positions and the character of shear stress alterations in time for clinker pastes C_2, C_3 and also D_1, D_2 has proved that clinkers originating from the same cement plant reveal a different reactivity at early hydration stages. The fact that clinkers originating from the same source are characterized by a similar mineral composition and contain relatively small and approximate quantities of alkalies of comparable sulphatization degree indicates that differences found in the reactivity of these clinkers can not be explained to arise due to the alkali presence only. For these pastes also the regularity is maintained according to which the 28-day strengths are greater for cement pastes characterized by a lower consistency at the initial hydration stage, provided the w/c ratio is kept constant.

Results obtained in rheological tests of cement pastes containing alkalies predominantly as sulphates (sulphatization degree of alkalies greater than 75 %) are characterized by a low consistency and slight time dependence (Table 2). This is in good agreement with results reported by Richartz (1986).

Rising up the sulphatization degree when the content of K_2O is high in clinker reduces the amount of potassium incorporated into the C_3A, thus less influencing the increase

in reactivity of that phase. This is confirmed by microscopic studies and microcalorimetry which show the hydration
progress to be efficiently controled by gypsum. The fine
crystalline ettryngite precipitated on the C_3A surface
inhibits the development of the hydration process and is
responsible for the low paste´s consistency.

Clinker pastes prepared from materials which contained
minor alkali quantities of lower sulphatization degree can
be classified as less reactive ones. The consistency of cement pastes obtained from these clinkers and its time
dependence indicate for an improved flowability caused by
the addition of gypsum.

Examination of the rheological parameters of cement pastes including less quantities of alkalies built-in into the
clinker phases has proved the consistency of pastes not to
be affected considerably. The pastes itself were characterized by a low consistency. Thus, the effect of alkalies
towards increasing the cement paste consistency during the
initial hydration stages is found to be most significant
at larger alkali contents (greater then 0.80% wt Na_2O_e).

4 Conclusions

Rheological studies of cement pastes originating from the
source which contained approximate amounts of C_3A but
differing in type and quantity of alkaline compounds
allowed to specify the role of alkalies in modifying of
rheological properties.

In cement pastes exhibiting higher reactivity of the C_3A
phase due to the incorporated alkalies the efficiency of
control over the hydration process by gypsum is decreased.
It has been found the higher the reactivity of clinker due
to the increased reactivity of C_3A caused by alkalies
(potassium) the lesser is the efficiency of gypsum as a
setting control agent, this is exhibited by rapid raise of
the consistency. The most significant factor which
determines the rheology of these pastes are the alkalies
incorporated into the clinker phases. Contribution of the
C_3A phase is found to be less significant.

A definite relationship has been found between the
initial hydration stage consistency of cement pastes and
28-day strength. It has been proved that the highest
strengths were achieved by cement pastes which exhibited
the lowest consistency during the initial hydration stage.

5 References

Banfill, P.F.G. (1981a) On the viscometric examination of cement pastes. Cem. Concr. Res., 11, 363-370.

Banfill, P.F.G. (1981b) A viscometric study of cement pastes containing superplasticisers with a note on experimental techniques. Mag. Concr. Res., 33, 37-47.

Berg, vom W. (1979) Influence of specific surface and concentration of solid upon the flow behaviour of cement pastes. Mag. Concr. Res., 31, 211-216.

Bombled, J.P. (1980) Influence of sulphates on the rheological behaviour of cement pastes and their evolution. Proc. of 7th Int. Congress on the Chemistry of Cement, Editions Septima, vol.III, IV, Paris, 164-169.

Collepardi, M. (1971) The rheological behaviour of cement pastes. Il Cemento, 68, 99-106.

Grzeszczyk, S. (1989) **Rheological Properties of Cement Pastes and Reactivity of Clinker.** Monograph (in Polish), Technical University of Opole Press, Opole, Poland.

Grzeszczyk, S. and Kucharska, L.(1990) Hydrative reactivity of cement and rheological properties of fresh cement pastes. Cem. Concr. Res.,(in print).

Odler, I., Becker, T. and Weiss, B. (1978) Rheological properties of cement pastes. Il Cemento, 3, 303-310.

Richartz, W. (1986) Einfluss des K_2O-Gehalts und des Sulfatisierungsgrads auf Erstarfen und Erharten des Zements. **Zement-Kalk-Gips**, 39, 678-687.

INTERACTIONS BETWEEN SULPHATE MINERALS AND C$_3$A IN CEMENT PASTE RHEOLOGY

S. SUHR
Institut für Baustoffkunde und Materialprüfung, Universität
Hannover, Federal Republic of Germany

1   Introduction

The rheological properties of cement pastes are of interest
in the field of grouting. They also influence the consistency of fresh concretes and consequently their workability. The flow behaviour of the cement pastes is influenced
by temperature and by chemical and mineralogical composition of the cement. Especially strong influence is exerci-
sed on the properties at the fresh phase by the matching of
type and amount of calcium sulphate to the aluminate clinker components.
   A research programme was carried out at the Institute
for Building and Material Testing of the University of Hannover in order to investigate the reciprocal effects of
calcium sulphate modification and C$_3$A content of cement on
the rheological properties of cement pastes.

2   Original Materials

The cements tested were produced from four different portland cement clinkers. The clinkers originated from industrial production. Table 1 gives an abstract of the chemical
analysis and the mineral contents of the clinkers according
to Bogue. The C$_3$A content lay between calculated 0 % by
weight at D clinker and 10 % by weight. The clinkers were
ground to a fineness of 4500 cm^2/g according to Blaine. The
total sulphate content of the cements was 3 % by weight in
all cases. For this purpose, calcium sulphate was added to
the clinker, as a mixture of hemihydrate and anhydrite. The
proportion of hemihydrate and anhydrite of the cements of
each clinker was varied in six regular stages between 100 %
by weight of hemihydrate and 100 % by weight of anhydrite.
   It was assumed that, due to the higher solubility of the
hemihydrate as compared with the anhydrite, there were unequal sulphate amounts clearly present in the mixing water.
The available sulphate was thus varied for all clinkers
from a sulphate deficiency to a sulphate excess.

Table 1 Chemical and mineralogical components
          of the clinkers

type of component	part by weight as percentage			
	D	E	F	G
SiO_2	21.0	21.2	21.9	20.9
Al_2O_3	3.6	5.2	4.1	5.5
Fe_2O_3	7.7	2.1	2.7	2.9
CaO	64.6	67.6	67.2	65.9
SO_3	0.5	0.6	0.6	0.9
C_3S	64.1	75.9	75.3	68.8
C_2S	11.9	3.7	6.1	7.9
C_3A	–	9.9	6.4	9.6
C_4AF	17.0	6.7	8.1	9.0

3 Test programme and test method

The rheological properties of cement pastes were determined, plotting continuous flow curves with a rotary viscosimeter. In this way it is possible to determine the relationship between shear stress and shear rate under physically defined conditions. Flow curves are necessary in order to characterize cement pastes adequately as representa-
tives of non-Newtonian fluids, since their viscosity depends both on shear rate and on duration of loading /1/.
  The measurements were carried out using a temperable
Searle rotary viscosimeter made by Haake, Type Rotovisco RV
12. The programmer (PG 142) was also connected. Fig. 1
shows the measuring system used in cross-section. It consists of two coaxially spaced cylinders, which are grooved

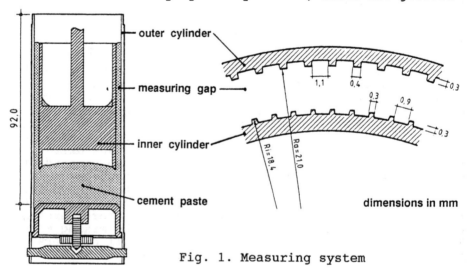

Fig. 1. Measuring system

in the same direction as the axis. This profile prevents the cement paste from demixing and improves adhesion in the periphery. An air cushion on the lower face of the inner cylinder guarantees that only the cement paste in the measuring gap between inner and outer cylinder is used for the test results. An outlet at the top of the cylinders allows excess test substance to run out.

When the inner cylinder turns, thin rotation surfaces are produced between them. The shear rate between these imaginary rotation surfaces is proportional to the turning speed. The shear stress resulting on the wall of the inner cylinder is recorded as a moment using a spring gauge.

The flow curves of cement pastes were determined as shown in fig. 2. Within 2 minutes, the turning speed of inner cylinder and, at the same time, the shear rate in the cement paste, were constantly increased up to a maximum of 512 rpm and then immediately slowed down to nil. During this procedure the shear stress produced was continuously recorded in a personal computer for later analysis. Parallel to this the flow curves were plotted as an optical control of the test procedure.

The cement pastes were made in a mixer according to DIN 1164 with a water/cement ratio of 0.44. The tests were carried out at temperatures of 5 °C, 20 °C and 30 °C after hydration periods of 10 and 30 minutes, each time with a new sample.

4   Results

4.1 Flow curves of cement pastes
The flow curves shown in the following illustrations are typical examples of the flow curves plotted under the chosen circumstances. At the same time they have the task of illustrating the effect of modification of the calcium sulphates on the one hand and of the C_3A content on the other on the whole shape of the flow curve.

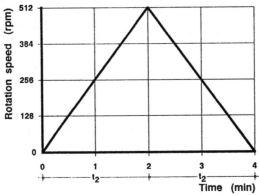

Fig. 2. Rotation speed of inner cylinder

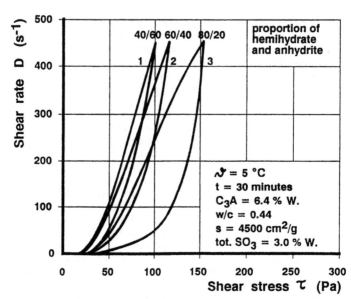

Fig. 3. Flow curves of cement pastes influenced
by the type of calcium sulphate

The flow curves shown in fig. 3 represent the result of measurements taken from cement pastes, produced from a cement with a C_3A content of 6.4 % by weight. At this stage we must remind ourselves that all research was carried out using cements with a specific surface of 4500 cm²/g and a total sulphate content of 3 % per weight. Therefore these factors will not be mentioned again. The x-axis represents the shear stress (τ) in Pascal units and the y-axis the shear rate (D) in units of s^{-1}. The flow curves were registrated 30 minutes after the start of mixing at a temperature of 5 °C. The differences between these flow curves is due to a change in the type of sulphate. From flow curve 1 to flow curve 3, the proportion of hemihydrate added is increased from 40 % to 80 % by weight.

While the same amount and type of sulphate is present, the difference in the flow curves shown in fig. 4 lies in the C_3A content of the cements. The flow curves were plotted at a temperature of 20 °C after a hydration period of 10 minutes. The C_3A content of the cements is increased from flow curve 4 to flow curve 7 by from 0 % up to 9.9 % by weight. However, it was not the cement with the highest C_3A content (curve 4) which produced the thickest paste, but, instead, the cement with the highest alumina content (curve 3).

As can be seen from the illustrated flow curves (and indeed from the flow curves of nearly all the cement pastes tested), cement pastes can be described as non-Newtonian fluids with thixotropic behaviour and structural viscosity.

40

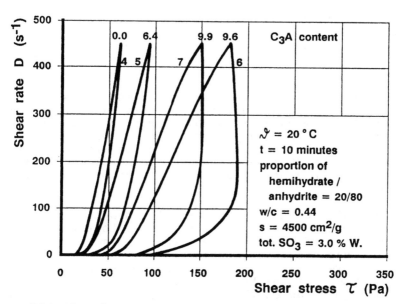

Fig. 4. Flow curves of cement pastes influenced
by the C₃A content

The loading curves of those cement pastes tested are at
first strongly curved, becoming steadily less curved and
even straight, as the shear rate increases. A major part of
the shear thinning ends roughly in the first quarter of the
measured area of shear rate. Once the maximum shear rate is
reached, the load curve is followed by the deloading curve,
which is on the whole more straight.

The original viscosity can only be regained -if at all-
by retard. The structural formation in the cement paste becomes reduced or even destroyed by mechanical stress. The
loading and deloading branches of the flow curve therefore
outline a hysteresis area, typical for thixotropic substances. This hysteresis bulge, whose area represents the work
related to the sheered volume, allows us a qualitative
judgment of the shear load brought in for the sake of gau-
ging, thus allowing conclusions as to the stability of the
cement paste structure.

This qualitative description of the types of flow curve,
determined during rheological research on cement pastes,
will have to be adequate, since the effect on flow behaviour is the main point in question.

In order to define the flow behaviour of the cement
pastes, the following rheological values were selected from
the information stored:

Yield point τ_0, which is the lowest shear stress measured at the start of the shear loading.
Initial viscosity η_i, calculated from the cotangent of

the angle between x-axis and a line cutting the primary flow curve.
Thixotropic energy A_h, which is the work done during the shear test related to volume, being the area of the hysteresis bulge, multiplied by the duration of loading.

It was noticed that the thixotropic energy is proportional to the yield point, although it only takes the primary stage of the flow curve into consideration. Therefore this value will not be used any further. Also, only the measurements taken after 10 minutes of hydration are to be presented, since after 30 minutes the values merely increased, without appearence of any different relationship to the influences.

4.2 Effect of type and amount of calcium sulphate
Earlier research at the Institute showed that the rheological properties of cement pastes are influenced by type and amount of sulphate in the cement /3/.
The measured values showed a dependence on the structural development of the hydration products. The structural development depends on the solubility of calcium sulphate and on the reactivity of C_3A, which are again effected by temperature. By harmonizing the sulphate content with the reacting C_3A, the greatest possible delay of the structural development can be achieved /2/. This minimizes the yield point because only ettringit is formed.
Fig. 5 gives an example of the yield point (τ_0) of cements from F clinker with a C_3A content of 6.4 % by weight, demonstrating the relationship to the SO_3 content of the hemihydrate added. The lowest yield point occurred with a hemihydrate value of about 1.0 % by weight of SO_3. Both lower and higher hemihydrate contents caused a rise in the yield point due to additional reaction products. This was caused in the left-hand part of the graph by a lack of easily soluble hemihydrate. In the area to the right, the increase was due to a tendency to false stiffening, caused by a high proportioning of hemihydrate in the cement.
The above mentioned systematic nature of the dependency of yield point on hemihydrate content was similar in the cases of all clinkers tested. The influence of the anhydrite was graded as small. Only where nothing but anhydrite is present in the cement, an influence can be determined compared with clinker without sulphate addition.
The influence of temperature on the reactivity of C_3A was demonstrated in the way the yield point increased as a result of increased reaction products. With increasing temperature a minimum was achieved at a higher level.
Fig. 6 shows, however, that a turn may occur in the dependency of the yield point on the temperature. This could be observed in the case of G clinker with the highest aluminium oxide content.
As opposed to the yield point, the initial viscosity

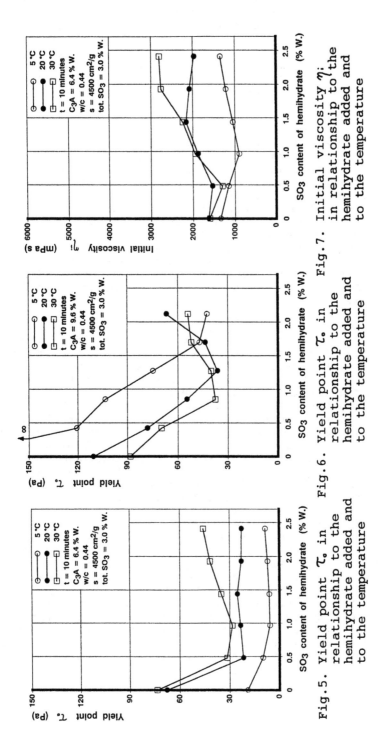

Fig.5. Yield point τ_o in relationship to the hemihydrate added and to the temperature

Fig.6. Yield point τ_o in relationship to the hemihydrate added and to the temperature

Fig.7. Initial viscosity η_i in relationship to the hemihydrate added and to the temperature

shows a principally different dependency on sulphate content and temperature. Fig. 7 shows the initial viscosity corresponding to the previously shown yield points of F clinker, once again depending on the SO_3 content of the hemihydrate added. There is a tendency for the cement pastes to thicken as the SO_3 content increases and the temperature rises. This also applies to the other clinker tested, with the exception of clinkers with high C_3A content at low temperatures.

The influence of the sulphate content on the yield point and initial viscosity can roughly be described as follows: excess sulphate causes a raising of the yield point and the initial viscosity by recristallization of the calcium sulphate. Sulphate deficiency however raises the yield point but lowers the initial viscosity, because the development of monosulphate increases, while that of the ettringit decreases. Here the yield point and the initial viscosity react independantly of each other. Thus, in order to compare rheological properties of various cement pastes, whose flow behaviour was determined under identical circumstances, it is necessary to give both factors. The thixotropic energy factor may be added in cases where the yield point and the initial viscosity do not allow an adequate differentation.

4.3 Influence of C_3A content
A distinct influence of the C_3A content was determined. It was proved, that the total C_3A content is not responsible for the reactivity of the C_3A. The decisive point is the proportion reacting immediately respectively the amount of C_3A reacting in the initial minutes.

Fig. 8 shows the yield point of the cement pastes of all clinkers in relationship to the SO_3 content of clinker and hemihydrate and to the C_3A content. These values were measured at a temperature of 20 °C and after 10 minutes hydration. Especially where the sulphate content was comparativelly low, the yield point of the cement pastes increased considerably. The cement pastes of G clinker give the highest yield point, which mean that we can assume that here the highest proportion of C_3A is reacting, although E clinker contains most C_3A.

As the C_3A content increases, the sulphate content, which minimizes the yield point, grows. It may be clearly seen that cements with lower C_3A content are less influenced by modification of the calcium sulphate. It was also noticed that the sensitivity to the available sulphate of these clinkers differs. Whereas an easily recognizable minimum is determined with G clinker, the other clinkers react less lively to changes in the available sulphate. The yield point of these clinkers also increases very little as the sulphate content is raised, while the sharp rise in the yield point of G clinker leads one to assume an excess of sulphate.

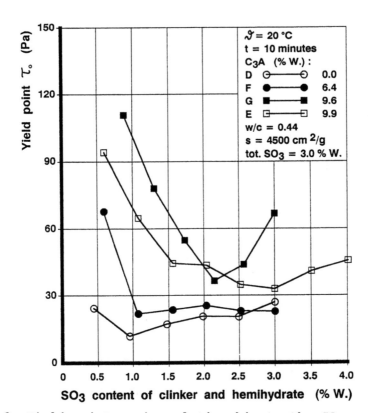

Fig. 8. Yield point τ_0 in relationship to the SO_3 content of clinker and hemihydrate and to the C_3A content

Just as in the case of the yield point, an increase of initial viscosity is seen to accompany the rising C_3A content. Fig. 9 shows the initial viscosity η_i measured together with yield points shown before, in relationship to the SO_3 content of clinker and hemihydrate and to the C_3A content. In contrast to the yield point, the influence of C_3A remains, even with increased sulphate content. This may be seen from the roughly parallel increase of the initial viscosity of all clinkers under increasing sulphate content.

5   Final remarks

We can conclude by stating that by recording flow curves with a rotary viscosimeter and by evaluating the results along with the factors yield point and initial viscosity, the influences on the rheological properties of cement pastes may be accurately measured. Decisive for the flow behaviour is the degree of coherence and movability of the cement particles, as well as the hydration products or

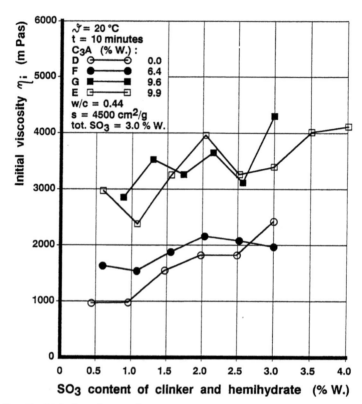

SO3 content of clinker and hemihydrate (% W.)

Fig. 9. Initial viscosity η_i in relationship to the
SO₃ content of clinker and hemihydrate and to the
C₃A content

group of hydration products, depending on the sulphate
available, the proportion of C₃A reacting immediately and
the temperature.

6 References

/1/ Berg, W. vom, Zum Fließverhalten von Zementsuspensionen. Dissertation, RWTH Aachen (1982).
/2/ Locher, F. W., Richartz, W., Sprung, S., Erstarren von
Zement. Teil 1: Reaktion und Gefügeentwicklung. Zement-Kalk-Gips 29 (1976) Nr. 10, 435-442. Teil 2:
Einfluß des Calciumsulfatzusatzes. Zement-Kalk-Gips
33 (1980) Nr. 6, 271-277.
/3/ Suhr, S., Einfluß von Calciumsulfaten im Zement auf die
Rheologie von Zementleimen, in Berichte vom Fachkol-
loquium "Zementleim, Frischmörtel, Frischbeton", Ei-
genverlag des Institutes für Baustoffkunde und Materialprüfung der Universität Hannover (1987) Heft 55.

6 EFFECT OF PRODUCTS OF THE HYDRATION OF C₃A ON RHEOLOGY OF CLINKER AND CEMENT PASTES

L. KUCHARSKA
Institute of Building Science, Wroclaw Technical University, Poland

Abstract
Flow curves, yield stress and plastic viscosity of fresh clinker and cement pastes with and without hydration products were determined from coaxial cylinder viscometer measurements. These were related to kind of hydration product with regard to hydraulic reactivity of aluminate phase and availibility of sulphates.
Keywords: Rheology, Cement Pastes, Hydration Products, Clinker Pastes, Flow Curves, Yield Value, Plastic Viscosity.

1 Introduction

It is reasonable to assume that the principal products of early hydration of portland cement are formed by the reaction of aluminate phase, which is manifested by a change in the rheological properties of the fresh cement pastes [Ish-Shalom and Greenberg (1962), Banfill and Saunders (1983), Caufin and Papo (1984), Uchikawa et al.(1985)].

According to Locher (1980), Hampson and Bailey (1983), Bombled (1980), the quantity of product hydration is affected by the amount and hydraulic reactivity of C₃A phase whereas the product kind by the liquid phase composition (SO₄²⁻ , pH).

It is generally accepted that in the first stage a small quantity of C₃A reacts. As Locher (1980) showed, the quantity of C₃A reacted in clinker and cement pastes are similar. Great differences in rheological behaviour of pastes without and with gypsum are connected mainly with the hydration product kind and its influence on grain surface modification (intergranular interference forces, coagulation structure formation, physical immobilization of water etc.). From among several hydration products which can form and their mixtures [Hampson and Bailey (1982)(1983), Frigione (1983)], the most advantageous for rheological properties of cement pastes is fine-grained crystal of ettryngite [Locher et al.(1980)]. Variations in reactivity of C₃A phase in industrial cements, even within these coming from the same cement plant result in the fact that conditions to form only fine-crystalline ettryngite deviate more or less from optimal ones, which is manifested by their various rheological behaviour [Odler and Abdul-Maula (1984), Uchikawa (1985)].

A great number of factors which differentiate rheological properties of cement pastes hinders on indirect evolution of hydration products of C_3A phase, which differentiate rheological properties of cement pastes. Assuming that the aluminate phase hydration includes the formation of diffusion barriers [Birchall et al.(1980), Tadros et al.(1976)] it is possible to evaluate the influence of crystallizing hydrates coming from the solution (ettryngite, monosulphate, hydrous calcium aluminate) by limiting the hydration to the topochemically formed layer.

According to Birchall et al.(1978), if pH of the liquid phase is supressed, the aluminate in portland cement (as with the silicates) does not react. This also appears to be true for C_3A and C_3A plus gypsum [Tadros et al.(1976), Skalny and Tadros (1977)]. As Lieber (1981) showed, an admixture of H_3BO_3 to cement pastes retards the bonding of sulphates and eliminates the initial phase of ettryngite formation. From the paper of Kucharska and Grzeszczyk (1989) it follows that in the presence of H_3BO_3 also the hydration of clinkers and pure C_3A phase without and with gypsum is retarded.

This paper presents the results of indirect evaluating the influence of the products of aluminate hydration phase on rheological behaviour of clinker and cement pastes with regard to the influence of hydraulic reactivity of C_3A and the demand for sulphates resulting from it.

2 Experimental

2.1 Materials
Portland cement clinkers, industrial (K_1-K_5, Table 1) as well as laboratory synthesized (K_6,K_7) of composition as given in Table 1, and corresponding cements (C_1-C_7) were ground to SSB = 300 $\pm$ 10 m^2/kg. Irrespective of contents and reactivity of C_3A phase, the cements were prepared using 5% naturaldihydrate calcium sulphate (2.2% w/w SO_3).

The pastes were made by mixing the clinker or the cement with deionised water or with a solution of 2.5% w/w boric acid. The water/cement ration was 0.5 for the clinker pastes without addition and 0.4 for clinker and cement pastes with and without boric acid. The conditions for preparing the pastes (mode of mixing, intensity, time, temperature about 20°C) were the same for all the samples examined. No special steps were taken to remove air from the mix.

2.2 Testing procedure
The tests were performed with a rotating coaxial cylinder viscometer Rheotes RV-2 using device H. Measureble stress range was 2.64-2588 Pa. The shear cycle started 10 minutes from the first addition of water or acid solution. The time for a complete increase-decrease cycle was 12-15 minutes.

Table 1. Characteristics of clinker (K) and surface area of cement (C)

	K_1	K_2	K_3	K_4	K_5	K_6	K_7
C_3S	36.0	57.8	50.0	58.6	53.0	49.7	50.3
C_2S	37.0	20.4	26.0	14.9	20.2	24.0	23.5
C_3A	4.7	8.5	8.6	9.3	15.3	14.4	14.0
C_4AF	16.7	8.8	9.3	13.4	7.6	11.3	11.7
Na_2O	0.1	0.2	0.2	0.1	0.2	0.8	0.0
K_2O	0.9	0.5	1.3	0.7	1.2	0.0	0.8
SO_3	1.0	0.9	1.5	0.7	0.7	0.0	0.0
Frec CaO	0.8	1.2	0.7	1.0	1.2	0.8	0.8
SSB m^2/kg	301	310	296	303	299	302	301

	C_1	C_2	C_3	C_4	C_5	C_6	C_7
SSB m^2/kg	303	303	308	303	301	303	301

3 Results and discussion

The investigation of the effect of products hydration on flow curves of clinker pastes (Fig.1) has shown that the hydraulic reactivity and the contents of aluminate phase has a considerable influence on flow kind and the Bingham parameters (Tab.2).

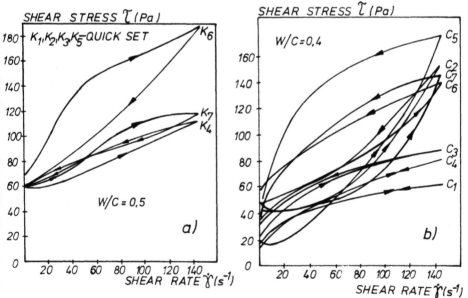

Fig.1. The flow curves of clinker (a) and cement (b) pastes with hydration products

Hysteresis is characteristic of clinker pastes free from sulphates. More reactive is the clinker (K_6), the greater it is. The occurrence of anti-loop on the flow curves of K_4 (Fig.1 a) clinker slurry (0.7% SO_3) indicates the formation of aluminosulphates, apart of gell like hydrate calcium aluminates. The disadvantageous influence of hydration products mixture results in a quick set of K_1, K_2, K_3 and K_5 clinker pastes. These clinkers contain 1, 0.9, 1.5 and 0.7% w/w of SO_3 respectively (Tab.1).

With regard to dihydrate calcium sulphate of the same quantity and grinding, the flow curves of cement pastes (Fig.1 b) reveal the influence of hydration product kind correspondingly to the clinker reactivity.

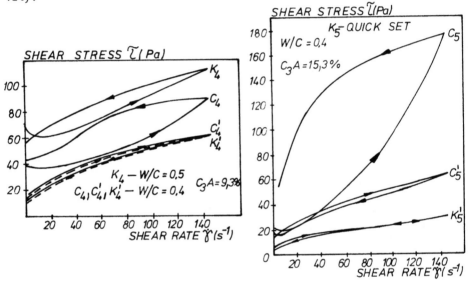

Fig.2. Flow curves of clinker and cement pastes with and without hydration products.

An examination of Fig.1 b shows that with regard to the calcium sulphate used the conditions are fulfilled for forming stabilizing properties of ettryngite in cements of low C_3A contents (C_1) and of higher contents of C_3A and lower reactivity (C_3, C_4). The Bingham flow is characteristic of these pastes. The formation of more than one hydration products kind, if the supply of sulphate is inadequate for quantity and reactivity of aluminate phase, corresponds with the occurrence of a large anti-loop on flow curves and the Bingham parameters (yield value, plastic viscosity, see Tab.2) increase. The elimination of hydration products drastically decreases the consistency of clinker pastes (Fig.2, curves K, K´, a decrease of w/c by 0.1), and to a lesser degree, that of cement pastes (Fig.2, curves C, C´) independently of C_3A contents.

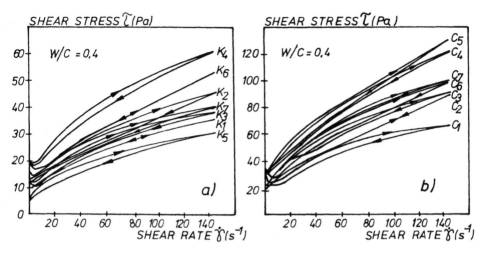

Fig.3. Flow curves of clinker (a) and cement (b) pastes
without hydration products

From the analysis of Fig.3 a and 3 b it results that all pastes
examined are characterized by a flow approximating that of Bingham,
with a small hysteresis, probably resulting from a slight sedimentation during the measurement. As it results from Fig.3, the elimination
of hydration products by hydration retarding through adding boric acid
practically eliminates the difference in rheological behaviour of clinker and cement pastes. The small evident differences in the position
of flow curves may be more the result of incontrollable factors connected with synthesis, grinding, pastes preparing, the measurement itself
than the influence of clinker composition or sulphates added.

Table 2. Bingham parameters of the clinker and cement pastes with
(K,C) and without (K´,C´) of the hydration product

	Yield value (Pa)				Plastic viscosity (Pa.s)			
	τ_K	$\tau_{K'}$	τ_C	$\tau_{C'}$	η_K	$\eta_{K'}$	η_C	$\eta_{C'}$
K_1, C_1		13	30	13.5		0.13	0.30	0.17
K_2, C_2		12	32	12.0		0.16	0.50	0.22
K_3, C_3		12	40	14.0		0.20	0.43	0.23
K_4, C_4	63	22	27	20	0.40	0.30	0.35	0.30
K_5, C_5		8.5	112	16		0.11	0.55	0.32
K_6, C_6	58	17	69	13	0.80	0.26	0.57	0.25
K_7, C_7	66	15	68	20	0.62	0.17	0.60	0.16

|←w/c = 0.5→|← w/c = 0.4→|←w/c=0.5→|← w/c = 0.4 →|

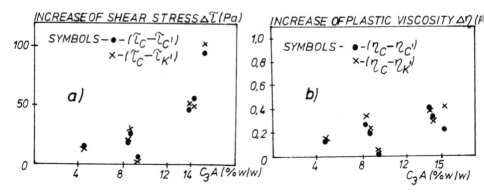

Fig.4. Influence of C₃A contents on increase in yield value (a)
and plastic viscosity (b) of clinker and cement pastes
by hydration product.

As it follows from Tab.2 and Fig.4 the effect of hydration products
on yield value is stronger on plastic viscosity and its affected by
C_3A phase contents and its reactivity. It increases as the availibility
of sulphates for the formation of fine-crystalline ettryngite decreases
    From among 7 cement pastes tested the optimal conditions appear
only in one cement paste (C_4 - 9.3% C_3A). For this cement paste
hydration products have a little influence on Bingham parameter (Fig.4)
The clinkers K_6 (14.4% w/w C_3A), K_7 (14% w/w C_3A) and lesser K_7
(15.3% w/w C_3H) contain built-in alkalies (Tab.1). According to
Hampson and Bailey (1982) (1983), the control of the setting of portland cement by addition of gypsum is only possible in the pH rang ~
12.3 to 12.8. Alkalies released gradually from the network of minerals
influence the structure and permeability of the topochemically created
diffusion barrier, raising the pH above - 12.8, the make the conditions
for forming calcium aluminate monosulphate or other hydration products.
For these pastes antiloope increases correspondingly to the product
kind and quantity being formed (Fig.1).
    As it follows from the paper of Bombled (1980) the coagulating
effect of sulphate should occur in the flow curves and Bingham
parameters of cement pastes without hydration of products. The results
of Fig.3 b and Tab.2 do not confirm such action. It may be connected
with suppresed pH by additions H_3BO_3, to range approximates to the
izoelectric point of aluminium hydroxides. Under those conditions,
the sulphate ions do not have influence on a potential 𝒞 on grain
surface.

4 Conclusions

1. The more differentiated are the hydration products of aluminate phase, the greater is their influence on rheological properties of fresh cement pastes.
2. The hydration product kind of C₃A phase influences the flow curve type and Bingham parameters. If pastes are free of SO₃ it is manifested by "structural brake down" (histeresis) on flow curves. Under the conditions only of ettryngite formation, the flow approximates the Bingham flow. Under the conditions for forming mixed products, flow curves show an antiloop (rheopexy, antitixotropy).

5 References

Banfill, P.F.G. and Saunders,D.C. (1981) On the viscometric examination of cement pastes. **Cem.Concr.Res.**, 11, 3, 363-370.
Birchall,J.D., Howard,A.J. and Bailey,J.E. (1978) On the hydration of portland cement. **Proc.R.Soc.**, London, Ser.A, 360 (1702), 445-453
Birchall, J.D., Howard,A.J. and Double, D.D. (1980) Some general consideration of a membrane/osmosis model for portland cement hydration. **Cem.Concr.Res.**, 10, 2, 145-156.
Bombled,J.P. (1980) Influence of sulphates on the rheological behaviour of cement pastes. **Proc.7th ICCC. Ed.Septima,** III, VI, Paris, 164-169.
Caufin,B. and Papo,A. (1984) Rheological behaviour of cement pastes. **Zement-Kalk-Gips,** 12, 652-661.
Frigione,G. (1983) Gypsum in Cement, in **Advances in Cement Technology, Ed. S.N.Ghosh, Pergamon Press Ltd., Oxford, 485-536.**
Hampson,C.J. and Bailey,J.E. (1982) On the structure of some precipitated calcium aluminosulphate hydrates. J.Mat.Sci., 17, 3341-3346.
Hampson, C.J. and Bailey,J.E. (1983) The microstructure of the hydration products of tri-calcium aluminate in the presence of gypsum. **J.Mat.Sci.,** 18, 401-410.
Ish-Shalom, M and Greenberg, S.A. (1962) The rheology of fresh portland cement pastes, **Proc.4th Int.Symposium on the Chemistry of Cement,** Washington 1960, Monograf 43, 2, Nat.Bur.Stand. Washington D.C., 737-748.
Kucharska,L. and Grzeszczyk, S.(1989) Rheological properties and hydration of portland clinker and cement in the presence of polybasic acid. **Conference on Silicate Industry and Silicate Science,** Budapest, 12-16 June, 315-320.
Lieber,W. (1981) Einflus von anorganischen Borverbindungen aud des Erstarren und Erharten von Portlandzementen. **Zement-Kalk-Gips,**9, 473-475.
Locher,F.W., Richartz,W. and Sprung,S.(1980) Erstarren von Zement, Teil II, Einflus ds Calciumsufatzusatzes. **Zement-Kalk-Gips,** 6, 33, 271-277.
Locher,F.W. (1980) Hydration of Pure Portland Cements. **Proc.7th ICCC,** Ed.Septima, IV, VII, Paris, 49-55.
Odler, J. and Abdul-Maula,S. (1984) Über die Reaktivität von **Industrie-Klinkern. Zement-Kalk-Gips,** 6,37,311-315.

Skalny,J.Tadros,M.E.(1977) Retardation of tricalcium aluminate
hydration by sulphates. **J.Am.Ceram.60**, 174-175.
Tadros, M.E. Jackon,W.V. and Skalny, J.(1976) Study of the dissolu-
tion and electrokinetic behaviour of tricalcium aluminates.
Colloid and Interface Science, VI.M.Kerker, 211-213.
Uchikawa,H.Ogawa,K. and Uchida,S. (1985) Influence of caracter
clinker on the early hydration process and rheological property
of cement paste. **Cem.Concr.Res.**, 15, 4, 561-572.

THE RELATIONSHIP BETWEEN MICROSTRUCTURE AND RHEOLOGY OF FRESH CEMENT PASTES

X. YONGMO and H. DANENG
China Building Materials Academy, Beijing, China

Abstract
This paper analysed the occurring process of thixotropic curve, antithixotropic curve and interleaving curve of fresh cement pastes by means of step-increasing method. The relationship between microstructure and thixotropy of the pastes and the problem of unified rheological model for four kinds of stable flow curves were discussed.
Keywords: Cement paste, Step-increasing method, Thixotropy, Antithixotropy, Flow type, Model, Microstructure.

1 Introduction

The rheological behaviour of fresh cement pastes is the property of its microstructure. There are still, however, lots of problems in testing rheological behaviour, which hinder a further rheological study of fresh cement pastes. It is necessary and feasible to adopt unified instruments and mixing procedure to solve some of the problems. Testing techniques remain to be studied further. A new method should characterize the relationship between the microstructure state and rheological behaviour of cement pastes.

2 Experiment and results

The apparatus used in this work was a NXS-11 coaxial cylinder viscometer with outer cylinder radius 1.462 cm, inner cylinder radius 1.262 cm, height 5.0 cm. The thixotropy of cement pastes was determined by means of a step-increasing method and then down-curve was measured at constant speed. The down-curve displayed a stable flow curve and the time factor can be neglected in the curve measurement.

The effect of cycle time on thixotropic-antithixotropic interleaving behaviours was investigated and 5 forms of curves in 3 types were presented by Banfill and Saunders. But it was not clearly shown how 3 kinds of interleaving curves in the Type II produced [1]. In this paper, a

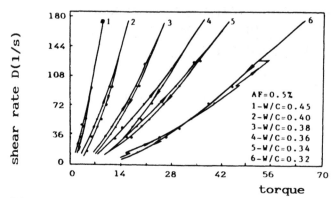

Fig.1 Antithixotropic behaviour with no structure breakdown

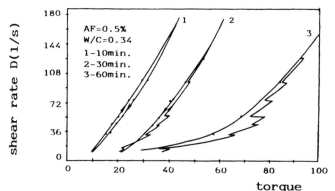

Fig.2 Thixotropic behaviour with structure
breakdown

step-increasing method was used to analyse the occurring
process of the curves.

Figure 1 presents the flow curves at different W/C
ratios when superplasticizer AF was 0.5%. There was no
thixotropic step shown on curve 1,2,3,4 or 5, this means
that the cement particles in slurry were entirely dispersive. Curve 1 approximated to reversible behaviour.
Antithixotropic behaviour became much more with the reduction of W/C ratio, this means that the effect of shear on
the hydration was related to the concentration of cement
pastes. Thixotropic breakdown occurred at higher shear
rate on curve 6 indicates that there were still non-
dispersed partical structure with higher bond energy in the
paste.

Figure 2 reports the effect of hydration on rheological
behaviour of the cement paste. After 30 minutes of hydration a floc structure was formed and strengthed with
hydration process. At 60 minutes of hydration, the

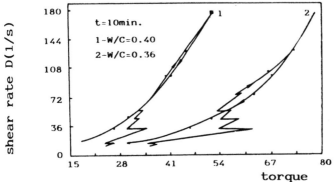

Fig.3   Anti-thixotropic behaviour (IIb) with
         structure breakdown

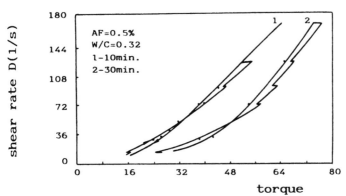

Fig.4   Anti-thixotropic behaviour (IIc) with
         structure beakdown

thixotropic step was greatly developed.

Figure 3 shows an antithixotropic behaviour (Type IIb) with thixotropic breakdown. The structure was entirely broken at lower shear rates, while thixotropic breakdown did not occur at subsequent higher shear rates. The antithixotropic behaviour occurred within higher shear rate range means only the effect of shear on the hydration of the cement.

Figure 4 reveals another structure state of cement pastes. No thixotropic breakdown existed at lower shear rates, while at higher shear rates, thixotropic step existed, resulting in the upper curve of thixotropic type and the lower curve of antithixotropic type, which is most frequent in the condition of water reducer addition and low W/C ratio. It is propably due to the fact that structure of hgiher bond energy which broke down only at higher shear rates was not thoroughly borken by mixing and water reducer.

3 Discussion

3.1 Interleaving behaviours and the structure of pastes
According to the results from the above the thixotropic
step was due to shear fracture of paste structure, while,
as a result of the effect of shear on the hydration,
antithixotropic behaviour was not appropriate as a parameter of paste structure. In the present work, kinds of
curves such as thixotropic type, reversible type, Type IIb,
Type IIc antithixotropic type and were obtained. It is
sufficient to explain how a complex interleaving behaviour
was caused due to shear testing method, structure state of
pastes and time of testing, though Type IIa curve did not
appear. The author advanced a new quantitative method of
the thixotropic steps and parameters that characterized
the paste structure [2].

3.2 Flow type and rheological model of fresh cement pastes
According to stable flow curve (down curve), the shear flow
of fresh cement pastes exibited Newtonian, Bingham plastic,
pseudoplastic and yield-pseudoplastic type. There should
be a unified rheological model that could describe the
four flow types under different conditions, and its each
parameter is with physical significance. The auther gave
recently a unified mathematical-physical model and a
rheological structure model [3,4].

4 Conclusion

The thixotropy and fluidity of fresh cement pastes are a
function of their structure state. The occurrence of
complex interleaving cycle curves is due to structure
states of the pastes, shearing method and time. Antithixotropic behaviour is caused by the effect of shear on
hydration and related to shear rate and solid-phase concentration. The different structure states of cement pastes
could be characterized by using step-increasing method
which is worth being perfected.

5 References

Banfill, P.F.G. and Saunders, D.C., (1981), **Cem. Concr.
Res.**, 11(3)363-370.
Xu Yongmo, Huang Daneng and Xie Yaosheng (1989), **J. Chinese
Ceramic Society**, 17(2)105-110.
Xu Yongmo, Huang Daneng (1989), **The 4th Asian Congr. of
Fluid Mechanics** (eds N.W.M. Ko and S.C. Kot), University
of Hong Kong, Vol.II, D39-D42.
Xu Yongmo (1989), **Proc. of 2nd Intern. Symp. on Cement and
Concrete**, (eds Yang Nanru, Jiang Jiafen and Huang
Shiyuan), Beijing, Vol.III, 172-178.

8 RHEOLOGICAL PROPERTIES AND AMOUNT OF EXPANSION OF LIGHTWEIGHT FRESH CEMENTS

C. ATZENI, L. MASSIDDA and U. SANNA
Dipartimento di Ingegneria Chimica e Materiali,
Università degli Studi di Cagliari, Cagliari, Italy

Abstract
The ability to take advantage of gas that develops in the
fresh mixes used for manufacturing lighweight cement
products, depends basically on their rheological properties.
Here the relationships between expansile properties, and
hence the resulting specific weights, and the rheological
parameters measured with a rotation viscometer are discussed
from a quantitative standpoint. The tested systems were made
up of both plain Portland cement and admixed with fly ash
(up to 60% by weight). The effect of acrylic latex and a
naphtalenesulphonate-based plasticizing admixture is also
examined.
Keywords: Lightweight gas concretes, Rheological properties,
Additives, Acrylic latices.

1 Introduction

The monitoring of the rheological properties of fresh
cements is of prime importance in cement technology (Bombled
1986, Tattersall and Banfill 1983). In particular for
lightweight aerated concretes produced using the technique
of hydrogen formation due to reaction between aluminum
powder and the alkaline constituents of cement mix, the
ability of the material to effective exploit the gas formed
within the mass, depends decisively on its rheological
properties.
The technology of producing lightweight aerated concrete,
achieved by gas forming within the still plastic mass, has
been industrially developed for several decades now (Short
and Kinniburgh 1963). Because of its good insulating
properties this type of concrete offers great potential. The
latest investigations have focussed in particular on
optimization mix compositions that incorporate industrial
by-products (Buttermore 1984, Hums et Alii 1984) and on the
effects of voids morphology on the mechanical(Bascoul et
Alii 1987) and thermal properties (Cabrillac and Perrin
1987).

59

2 Experimental methods

A '425' Portland cement (Italian Standards, similar to ASTM type II) as received and admixed with 20, 40 and 60% by weight of pulverized fly-ash (PFA) were used in the tests. Their chemical composition is shown in Table I. Mixes were

Table I. Chemical composition of Portland cement and PFA

Component		Portland cement	Pulverized fly ash
SiO_2	%	21.5	40.3
CaO	%	60.3	8.60
Fe_2O_3	%	2.40	4.60
Al_2O_3	%	5.10	36.3
MgO	%	2.65	1.49
K_2O	%	0.76	0.06
Na_2O	%	0.52	0.07
L.of I.	%	3.80	5.50

prepared with water/binder ratio (w/b) (binder = cement + PFA) of 0.40, 0.45 and 0.50. Tests were also conducted on mixes containing a latex, an acqueous dispersion with 50% by weight of an acrylic copolimer (polymer/binder (p/b) = 0.05 by weight) and for the system with w/b ratio of 0.40, a naphtalenesulphonate based plasticizer (p/b =0.005 and 0.02 by weight). The expansile process for developing hydrogen was activated by aluminum powder (0.2 g/100 g binder) and NaOH (1 g/100 g binder). The caustic soda was dissolved in the mix water whereas the aluminum powder was dispersed after the other ingredients had been mixed. Rheological tests were performed with a coaxial cylinder viscometer (Atzeni et Alii 1986). It was not possible to determine exactly when expansion began, but it may said that the phenomenon became visibly appreciable between 15 and 20 mins after the dispersion of the aluminum and concluded between 1 and 2 h later, the systems containing latex taking the longer. The tested materials were aged in ambient conditions (temperature: 25°C, relative humidity: 70%) for 28 days and cured for 24 h in saturated steam at 70°C and for 8 h with high pressure steam at 180°C. Prior to steam curing the specimens were conditioned for 48 h in ambient conditions. After ageing, 50 mm cubic specimens were tested for compressive strenghts.

Results and discussion

.1 Rheological properties

he rheological curves obtained (see for example Fig.1 and
) appear to display an initial steady portion that departs
rom the maximum shear stress value (τ_M), to reach a minimum
τ_m) after about 2-5 mins in the systems without latex and
fter 5-10 mins in those admixed with this additive. This
teady portion is followed by another tract, rising on
verage, but somewhat irregular, especially for the stiffer
ystems. The marked stiffening effect exerted by the PFA is
ikely due to the presence of unburnt coal (see Table I).
n any case the curves seem to be well differentiated, thus
lemonstrating that the rheological technique used is
atisfactory for discriminating among the different systems.
s can be seen the latex exerts a distinct plasticizing
ffect (Atzeni et Alii 1989). This results both in terms of
bsolute shear stress values, lower than those of the
orresponding systems where latex is absent, and in a
latter and longer minimum curve. The first phenomenon may
e explained by the fact that the latex, present in the form
f tiny micron and submicron spheres, presumably acts as a
ubricant, whereas in interpreting the second one should
onsider that the latex slows down all the surface phenomena
uch as hydration of the cement and reaction of the aluminum
owder with the alkaline solution, due to its filming, and
ence practically insulating effect of reactive phases
(Chandra and Flodin 1987, Afridi et Alii 1989). Without the
ddition of latex it is moreover impossible to obtain mixes
ith high PFA, unless large amount of water are used
(w/b=0.50).
or the initial steady portion of the flow curve, certain
ypical rheological parameters can be derived, after an
approach based on Tattersall equation (Tattersall 1955), in
particular maximum shear stress (τ_M) and the difference
etween maximum and minimum shear stress (Δ) (see Table II).
The resort to a 'traditional' plasticizing admixture such as
naphtalenesulphonate-based one proved unsuccessful with
these particular systems. As can be seen from the data shown
y way of example in Fig.3, the addition of such additive in
the amount of p/b=0.005 results in a pronounced stiffening of
the mixes. Despite producing fluid mixes, larger proportions
of naphtalenesulphonate (p/b=0.02) inhibit expansion of the
system. This behaviour is probably due to the interaction
between the functional groups of additive and the caustic
soda used to activate the expansion process.

3.2 Relationship between rheology and expansion

The specific weights of the aged products depend not only on
the expandibility of the fresh cements, and hence on their
rheological properties, but also on the proportions of the
mix ingredients, as their specific weights differ

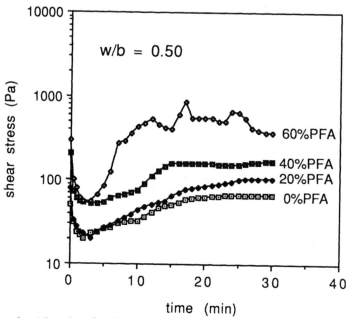

Fig. 1. Rheological curves:
shear stress vs time for the systems without latex.

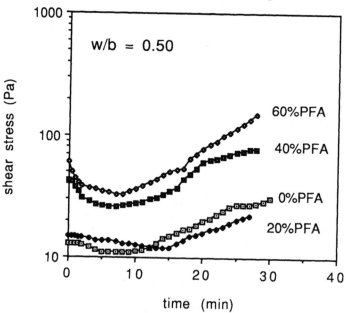

Fig. 2. Rheological curves:
shear stress vs time for the systems with latex.

Table II. Rheological properties (τ_M max shear stress, τ_m minimum shear stress, $\Delta = \tau_M - \tau_m$, Pa).

PFA % w/b		without latex				with latex			
		0	20	40	60	0	20	40	60
0.40	τ_M	85	230	1600	nw	43	130	240	500
	τ_m	32	58	650	nw	19	78	120	140
	Δ	53	172	950	nw	24	52	120	360
0.45	τ_M	70	200	500	nw	16	30	130	200
	τ_m	28	41	116	nw	15	25	63	76
	Δ	42	159	384	nw	1	5	67	144
0.50	τ_M	50	80	210	300	13	15	43	60
	τ_m	20	21	52	55	11	12	26	33
	Δ	30	59	158	245	2	3	17	27

nw = not workable.

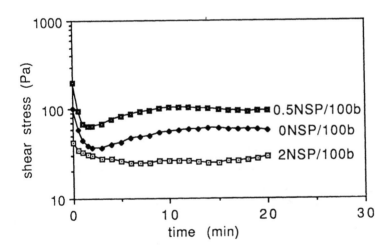

Fig. 3. Rheological curves:
Shear stress vs time for the systems with naphtalenesulphonate.

Table III. Experimental specific weights of the systems after 28 days ageing and reference specific weights (g/cm^3)

%PFA w/b	Without latex experimental				Without latex reference			
	0	20	40	60	0	20	40	60
0.40	0.87	1.09	1.14	nw	2.42	2.28	2.14	1.99
0.45	0.80	0.95	1.00	nw	2.38	2.21	2.10	1.96
0.50	0.75	0.86	0.87	1.03	2.33	2.20	2.06	1.93
	With latex experimental				With latex reference			
0.40	0.85	0.96	0.93	1.16	2.38	2.24	2.10	1.96
0.45	0.79	0.77	0.82	1.00	2.33	2.20	2.06	1.93
0.50	0.58	0.59	0.68	0.93	2.29	2.16	2.03	1.90

significantly (Portland cement: 3 g/cm^3; PFA: 2 g/cm^3; water and latex: 1 g/cm^3). Table III shows the specific weights of

Table IV. Parameter $-\Delta\rho$% for the different systems

%PFA w/b	Without latex				With latex			
	0	20	40	60	0	20	40	60
0.40	64.0	52.2	46.7	nw	64.3	57.1	55.7	41.0
0.45	66.4	57.5	52.4	nw	66.1	65.0	60.2	49.0
0.50	67.8	60.9	57.7	47.0	74.6	72.7	66.5	51.0

the specimens after 28 days ageing in ambient conditions and those calculated assuming a linear combination of the specific weights of the single ingredients, which can also be taken as a reference of the particular formulations. Table IV shows the parameter $-(\Delta\rho/\rho)$%, defined as the percentage variation of the experimental specific weight over that of the reference composition. Thus, this parameter conventionally takes into account only the system's

expandibility and hence any eventual correlations with the
rheological parameters defined in the previous paragraph,
can be examined. In Fig. 4 and 5 the parameter $-(\Delta\rho/\rho)\%$ is
plotted versus τ_M and Δ respectively. As can be seen within
the same w/b ratio expansivity depends on PFA content which
in turn affects shear stress values. All the systems show
sharp changes of this dependence for some values of both τ_M
and Δ. Expansion can therefore be optimized and controlled
to a certain extent by means of a rheological measure of
fresh mixes.

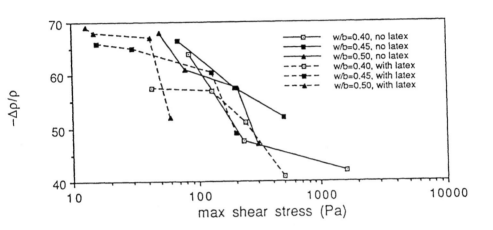

Fig.4 Parameter $-(\Delta\rho/\rho)\%$ vs τ_M (max shear stress).

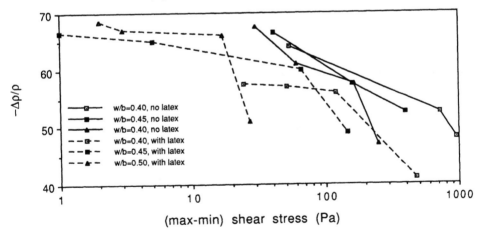

Fig.5 Parameter $-(\Delta\rho/\rho)\%$ vs Δ (max - min shear stress).

Table V Compressive strenght (S, MPa), specific weight (ρ, g/cm^3) and specific strenght (SS, MPa/(g/cm^3)) of the systems without latex.

w/b	%PFA	Curing 28 days 25°C				Steam curing 24h 70°C				Steam curing 8h 180°C			
		0	20	40	60	0	20	40	60	0	20	40	60
0.40	S	4.3	6.3	5.5	nw	3.3	4.2	5.9	nw	2.9	4.7	9.4	nw
	ρ	0.9	1.1	1.1	nw	0.8	0.9	1.0	nw	1.0	1.1	1.2	nw
	SS	4.9	5.8	4.9	nw	4.0	4.7	5.7	nw	3.0	4.1	7.8	nw
0.45	S	3.7	5.1	4.8	nw	2.4	2.9	3.5	nw	2.0	3.0	6.8	nw
	ρ	0.8	0.9	1.0	nw	0.7	0.8	0.9	nw	0.9	1.0	1.1	nw
	SS	4.6	5.4	4.8	nw	3.3	3.6	4.0	nw	2.3	3.1	6.3	nw
0.50	S	2.9	3.5	3.5	5.8	2.4	3.0	3.7	3.5	1.9	2.6	5.5	6.3
	ρ	0.7	0.9	0.9	1.0	0.7	0.7	0.8	1.0	1.0	1.1	1.2	1.1
	SS	3.9	4.1	4.0	5.6	3.6	4.0	4.6	3.6	1.9	2.3	4.7	5.8

Table VI Compressive strenght (S, MPa), specific weight (ρ, g/cm^3) and specific strenght (SS, MPa/(g/cm^3)) of the systems with latex.

w/b	%PFA	Curing 28 days 25°C				Steam curing 24h 70°C				Steam curing 8h 180°C			
		0	20	40	60	0	20	40	60	0	20	40	60
0.40	S	7.2	5.9	4.8	6.5	3.5	3.9	6.3	3.8	2.2	2.6	7.5	7.4
	ρ	1.0	1.0	0.9	1.2	0.7	0.7	1.0	1.1	1.0	1.0	1.2	1.3
	SS	7.1	6.2	5.2	5.6	5.0	5.2	6.9	3.4	2.2	2.5	6.3	5.7
0.45	S	3.6	3.3	3.0	4.6	3.3	2.1	2.7	3.1	1.8	1.6	3.5	5.0
	ρ	0.8	0.8	0.8	1.0	0.7	0.6	0.7	1.0	0.9	0.8	1.0	1.1
	SS	4.5	4.3	3.7	4.6	4.8	3.5	3.6	3.2	2.0	1.9	3.4	4.6
0.50	S	1.4	1.5	1.8	3.6	1.2	1.0	3.3	2.1	1.9	1.4	3.5	3.8
	ρ	0.7	0.6	0.7	0.9	0.7	0.7	0.8	0.8	0.9	0.8	1.0	1.0
	SS	2.0	2.6	2.7	3.9	1.8	1.3	4.1	2.7	2.2	1.7	3.6	3.9

.3 Mechanical properties

ompressive strenghts, specific weights and specific ompressive strenghts of the tested materials are shown in able V (systems containing no latex) and Table VI (systems dmixed with latex). As can be seen:
- the specific strenghts, especially of those specimens team cured at high and low temperatures, range, generally peaking, from 3 to 8 MPa/(g/cm^3);
- the addition of even large proportions of PFA imparts a marked improvement on mechanical properties, especially for specimens steam cured at 180°C;
- the addition of latex results, for equal w/b, in a decline n compressive strenght, as already reported in the iterature (Grosskurth and Konietzko 1987). Nevertless, expansive mixes can be obtained with these systems using ower water and high PFA (up to 60% by weight) contents, hat yield at the same time, good mechanical properties.

Conclusions

- The parameters τ_M and Δ, that can be derived from the transient measured with a rotation viscometer, correlate the expansibility of the different chemically aerated systems;
- high expansion can only be obtained for small values of the above parameters. The naphtalenesulphonate-based plasticizing admixtures normally employed are unsuitable for this purpose, wheareas the addition of polymeric latices permits to lower water content and use large proportions of PFA containing even considerable amounts of unburnt coal, as in the case at hand;
- the compressive strength values are relative to mix composition, specific weights and curing conditions; good mechanical properties have been obtained even for systems containing 60% PFA;
- rheological measurements are therefore useful in monitoring fresh mixes in lightweight concrete technology.

5 References

Afridi M.U.K., Ohama Y., Zafar Iqbal M. and Demuras K.(1989) Behaviour of Ca(OH)2 in polymer modified mortars.**The International Journal of Cement Composites and Lightweight Concrete**, 11,235-244.

Atzeni C., Massidda L. and Sanna U. (1986) Model for the thixotropic behavior of cement pastes. **Industrial and Engineering Chemistry,P.R. and D.**, 25, 499-504.

Atzeni C., Massidda L. and Sanna U. (1989) Rheological behaviour of cements mixed with polymeric latices. **The International Journal of Cement Composites and Lightweight Concrete**, 11,215-219.

Bascoul A., Cabrillac R. and Maso C. (1987) Influence de la forme et de l'orientation des pores sur le comportament mechaniques des betons cellulaires. In **Pore Structure and Construction Materials Properties** (Maso J.C. editor), Chapman and Hall, London, 333-340.

Bombled J.P. (1986) Influence des properties chimiques et physico-chimiques de la pate et des caracteristiques des granulats sur le comportement rheologique des mortiers et des betons. **Proceedings of 8th International Congress of Cement,** Rio de Janeiro, VI, 217-221.

Buttermore W.H. (1984) The development of lightweight concrete produced from pulverized coal fly ash and limestone fluidized bed combustor ash. **In Proceedings of Second International Conference on Ash Technology and Marketing,** London, 685-690.

Cabrillac R. and Perrin B. (1987) Influence de la forme et de l'orientation des pores sur la conductivite thermiques des milieux poreaux. In **Pore Structure and Construction Materials Properties** (Maso J.C. editor), Chapman and Hall, London, 341-348.

Chandra S. and Flodin P. (1987) Interactions of polymers and organic admixtures on portland cement hydration. **Cement and Concrete Research,** 17, 875-890.

Grosskurth K.P. and Konietzko A. (1987) Structure and mechanical behaviour of polymer modified cement concrete. In **Proceedings of 5th International Congress on Polymers in Concrete,** Brighton, 171-174.

Hums D., Hubert H. and Mortel (1984) Influence of different fly ash on the properties of aerated lightweight concrete. In **Proceedings of Second International Conference on Ash Technology and Marketing,** London, 685-690.

Short A. and Kinniburgh W. (1963) **Lightweight concrete.** C.R. Books Limited, London.

Tattersall G.H. (1955) The rheology of portland cement paste, **British Journal of Applied Physics,** 6, 165-167.

Tattersall G.H. and Banfill P.F.G. (1983) **The rheology of fresh concrete,** Pitmann, Boston.

9 RHEOLOGICAL MEASUREMENTS ON FRESH POLYMER-MODIFIED CEMENT PASTES

T. GREGORY and S.J. O'KEEFE
Bristol Polytechnic, Bristol, UK

Abstract
Rheological techniques have been used to measure the
behaviour of fresh cement pastes modified by the addition
of a stabilising surface active agent (SAA) and polymer
latex particles. The techniques were chosen to give
minimal disturbance of the age-dependent structure
development in the paste. The stabilising non-ionic SAA
used in the work was a nonylphenoxypoly(ethyleneoxy)
ethanol, and the latex was of the styrene-butadiene resin
(SBR) type. Two latex samples were used: one had a
particle diameter of 130 nm and the other a particle
diameter of 197 nm. All of the measurements were
performed on samples with a water:solids ratio of 0.39
(i.e. a volume fraction of solid particles = 0.45).
    The rheological techniques used were stress relaxation
after sudden strain, and constant frequency oscillation.
The decay of the stress was monitored and the relaxation
modulus calculated. In the oscillation experiments the
sample was subjected to a small amplitude, oscillatory
strain, and the storage and loss moduli evaluated from the
generated stress. The measurements in both sets of
experiments were performed as a function of the age of the
paste.
    An attempt is made to interpret the results at a
microscopic level, taking into account present theories on
the physical and chemical processes occurring in fresh
cement pastes.

1 Introduction

Rheological techniques have been used by many workers to
describe the time-dependent behaviour of modified and
unmodified fresh cement pastes.
    Banfill and Saunders (1981) used a viscometer with
coaxial cylinders geometry to measure the flow curves
(shear rate v. shear stress) of cement pastes from a shear
rate cycle. The hysteresis loops obtained were

interpreted in terms of structure breakdown (shear induced) and strength development (chemical hydration). Banfill (1981) studied the effect of the addition of superplasticising admixtures using different viscometers and geometries. The cement paste was assumed to behave as a Bingham plastic, and an apparent yield value and relative plastic viscosity evaluated. It was found that the former decreased as the dose of admixture increased, whilst the latter did not show an obvious trend as a function of admixture dosage. Lapasin et al. (1983) used different types of rheological measurement to probe the changes occurring physically (structural breakdown) and chemically (hydration) in fresh cement pastes. The effect of retarders (eg. sugars) on the rheological behaviour of fresh cement pastes has been studied by Bhatty and Banfill (1984). The apparent viscosity was monitored as a function of the age of the paste, and three distinct regions (designated I, II and III) were observed which were very sensitive to the concentration of the retarder. Region I showed an increase in the apparent viscosity, and was assumed to result from the reaggregation of cement particles after complete structural breakdown by mechanical shearing; region II was thought to correspond to a chemically dormant period since the apparent viscosity remained constant; and region III was assumed to reflect the onset of secondary hydration processes which caused an increase in the apparent viscosity. Hannant and Keating (1985) have reported the development of a new instrument to assess the strength development in fresh cement pastes by measurement of the shear modulus. They concluded that the new equipment allowed non-destructive measurements to be made on fragile suspensions. The data obtained also illustrated a slow change in the measured rheological parameter (due to physical processes) followed by a more rapid change (due to secondary hydration processes). Banfill and Gill (1986) have studied the steady shear rate viscosity behaviour of aluminous cement pastes. They observed a double minimum in a plot of torque against time of shearing. This was explained in terms of the flocculation/deflocculation processes occurring in the cement pastes. Gregory and O'Keefe (1988) reported some preliminary rheological data on polymer latex modified cement pastes. The measured rheological parameter was found to increase very slowly as a function of time before increasing by between 1 and 3 orders of magnitude over a relatively short time span. The time at which this transition occurred increased as a function of the dose of the polymer latex.

This paper describes the rheological behaviour of polymer latex modified cement pastes at a constant water : solids ratio. The techniques (stress relaxation and constant frequency oscillation) were chosen to give minimal disturbance of the age-dependent structure development in the paste. The data is interpreted using present theories describing the flocculation of colloidal dispersions and the hydration of cement particles.

2   Experimental Details

2.1 Materials and Sample Preparation

A special ordinary Portland cement (Wexham Developments Ltd., Slough) was used in the experiments. The material is prepared to BS12 for use with admixtures and when highly consistent material is required. The latices used were of the styrene-butadiene resin (SBR) type, which are stabilised during preparation by sodium dodecyl benzene sulphonate. Two latices were used in the work : one having a particle diameter of 197 nm and the other a particle diameter of 130 nm. The stabilising surface active agent (SAA) used in the study was a non-ionic nonylphenoxypoly(ethyleneoxy)ethanol with a relative molecular mass of 1540. The water used was distilled from an all pyrex glass system. All samples were prepared at a constant water : solids ratio of 0.39 (volume fraction of solid particles = 0.45). The cement powder was added to the water (which may contain SAA and SBR latex) and hand mixed for 30 seconds. This was followed by mechanical stirring using a Silverson L2R (Silverson Machines Ltd., Chesham, Bucks) for 90 seconds. The sample was then transferred to the concentric cylinders of the rheometer (maintained at a constant temperature of $20.0 \pm 0.1^{\circ}C$) for study.

The following samples were used in the study :

0	-	control paste ie. cement + water at a volume fraction of 0.45
0S	-	control paste with added SAA at 0.82×10^{-4} moles per 100g cement
10/1/5	-	control paste with 10% by volume of cement particles replaced by 197 nm latex particles and 7.27×10^{-5} moles SAA per 100g of solid
10/1/10	-	control paste with 10% by volume of cement particles replaced by 197 nm latex particles and 1.50×10^{-4} moles SAA per 100g of solid
10/2//5	-	control paste with 10% by volume of cement particles replaced by 130 nm latex particles and 7.27×10^{-5} moles SAA per 100g of solid
10/2/10	-	control paste with 10% by volume of cement particles replaced by 130 nm latex particles and 1.50×10^{-4} moles SAA per 100g of solid

2.2 Rheological Measurements

The instrument used to perform the rheological measurements was a Bohlin VOR Rheometer (Bohlin Reologi UK Ltd., Letchworth) with concentric cylinders geometry.

2.2.1 Stress relaxation

This test involves the rapid application of a small strain which is then held constant. The decay of the generated stress is monitored as a function of time. The test is schematically represented in Fig. 1. At any given time, t, the relaxation modulus, $G(t)$, is given by the ratio of the stress at that time, $\sigma(t)$, to the constant applied strain γ_o, ie. $G(t) = \sigma(t)/\gamma_o$. The initial modulus, G_{in}, is the modulus calculated when $t = 10$ ms after the strain ramp is complete, and the final modulus, G_{fn}, is the modulus when the system has substantially relaxed.

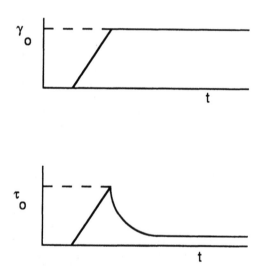

Fig. 1. Schematic representation of stress relaxation experiment

The sample was prepared as described in section 2.1 and then immediately introduced into the rheometer. A thin film of oil was floated on the sample to prevent drying during the experiment. Measurements were taken at regular intervals (every ten minutes) by application of the minimum strain which would produce a stress response measurable by the rheometer.

.2.2 Oscillation

his test involves the sample being subjected to a
inusoidally varying strain with measurement of the
esultant stress response. The measured stress is a
:omplex quantity and may be vectorially separated into two
:omponents : one component will be in phase with the
;train input, and the other component will be 90° out-of-
hase with the strain input. The stress component in
hase with the strain will reveal the storage modulus, G',
nd the stress component out-of-phase with the strain the
.oss modulus, G" (Ferry (1980)).

The sample was prepared for measurement as described in
;ection 2.2.1. Again measurements were taken at regular
intervals (about every ten minutes) at several frequencies
(from 0.005 to 20 Hz) using the same amplitude throughout
ie. 2 x 10^{-3} radians.

Results and Discussion

.1 Stress relaxation

Plots of the initial modulus, G_{in}, and the final modulus,
G_{fn}, against the age of the paste are shown in Fig. 2 (a)
and (b).

G initial (stress relaxation)

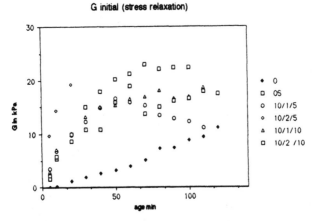

Fig. 2(a) Initial modulus, G_{in}, against age of paste

For the control paste the value of G_{in} increases by one
order of magnitude from approximately 1 kPa to 10 kPa over
a period of 130 minutes. For any of the modified samples
the value of G_{in} is much larger at any given age of the
paste. For all the modified samples, except 10/2/5, the
value of G_{in} increases by one order of magnitude over 40
minutes. For sample 10/2/5 the value of G_{in} increases at
an extremely rapid rate in the first 20 minutes after
mixing. Samples 0S, 10/1/5, 10/1/10 and 10/2/10 show

interesting behaviour at about 60-80 minutes when the
initial modulus plateaus or slightly decreases.

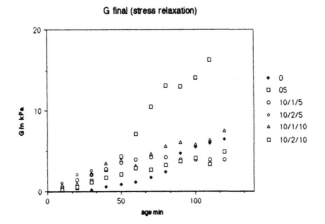

Fig. 2(b)   Final modulus G_{fn}, against age of paste

With respect to the final modulus, G_{fn}, the value
increases steadily over 120 minutes for all samples except
OS.   The mix with added SAA increases to 16 kPa after
about 100 minutes.

3.2 Oscillation
Plots of the data obtained from dynamic experiments at
0.005 Hz and 20 Hz are shown in Figs. 3 and 4,
respectively.   From the low frequency plots it can be seen
that at any given age of the paste the moduli magnitude
decreases in the order
0 > OS, 10/1/5, 10/2/5, 10/2/10 > 10/1/10.   For the
control paste the complex, storage and loss moduli
increase at a rapid rate after a short "dormant" period
(approximately 15 minutes).   The other mixes, except
10/1/10, exhibit a slow increase in moduli values.
Samples OS and 10/1/5 show anomolous loss modulus
behaviour since the magnitude starts to decrease at about
100 minutes.   Sample 10/1/10 has very low moduli values,
with only a very slight increase over 175 minutes.   The
high frequency data shows different behaviour since the
moduli magnitude decreases in the order 10/2/5 > 10/2/10,
10/1/5 > 0, OS, 10/1/10.   The shift of mixes 0 and OS to
smaller modulus values relative to other mixes, when
compared to the low frequency data is very noticeable.
Also the magnitudes of the moduli of these samples is seen
to increase at a rapid rate at about 75 minutes.   As with
the low frequency data the 10/1/10 sample shows a very

74

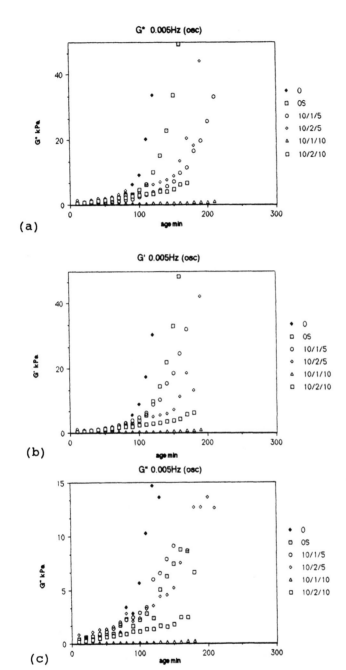

(a)

(b)

(c)

Fig. 3. Complex modulus (a), storage modulus (b) and loss
modulus (c) against age of paste at 0.005 Hz

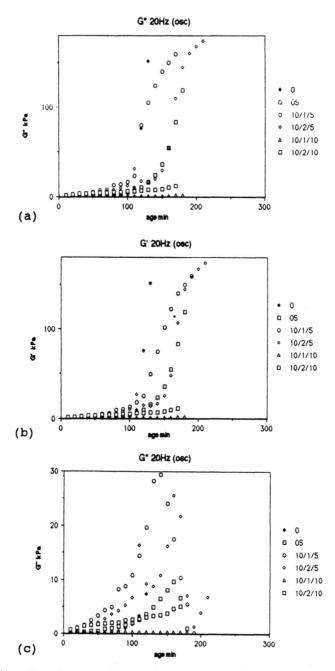

Fig. 4. Complex modulus (a), storage modulus (b) and loss
modulus (c) against age of paste at 20 Hz

shallow increase in the moduli over 175 minutes. The samples OS and 10/1/5 again show a peak value for the loss modulus.

3.3 Discussion

It is obvious from the relaxation and oscillation data that the experiments yield quite different information about the physical and chemical processes occurring in the cement pastes.

Prior to the rheological experiments comprehensive studies on the adsorption of the SAA on to both the cement particles and the latex particles were performed. The SAA level in sample OS represents a 10-fold excess of the SAA to completely saturate the surface of the cement particles. Therefore it may be assumed that (a) the cement particles will be completely dispersed in the mix, (b) a degree of multilayer adsorption may occur at the surfaces of the cement particles, and (c) there will be excess SAA in the aqueous phase of the mix. The concentration of SAA in (c) is in excess of the critical micelle concentration (cmc), and hence the SAA molecules will be present in the form of associations.

The levels of SAA in the mixes correspond to the 5% and 10% latex particles in the mix being partially stabilised (ie. incomplete coverage) or fully stabilised (ie. complete coverage), respectively. Indeed in the mixes containing 10% of SAA there will be sufficient excess SAA to completely disperse the cement particles.

The magnitude of the initial modulus in the stress relaxation experiments (Fig. 2(a)) is a measure of the solid-like properties of any given sample. As expected G_{in} increases steadily for the control paste (sample 0) as a function of its age. This is most probably due to the formation of a 3-D network of cement particles being formed in the paste. Over this time period the onset of the formation of secondary hydration products is not detectable, since it is assumed that this would cause the measured modulus to increase more rapidly. The sample containing 10% of the small latex particles and 5% of stabilising SAA (10/2/5) shows quite different behaviour since the measured modulus increases very rapidly over a short time span. It may be assumed that the SAA present in the mix will have a greater affinity for the hydrophobic material (ie. the latex particles) than for the hydrophilic material (ie. the cement particles). Thus the presence of these small latex particles appears to strengthen the flocculated network of cement particles. The mechanism for this is as yet unclear. The other mixes behave similarly and also show a plateau value for G_{in}, which indicates that a physical equilibrium has been obtained.

The magnitude of the final modulus (Fig. 2(b)) is a measure of the elastic nature of the "bonding" between the particles, or flocs, in the suspension. The anomolous behaviour of the OS sample after 50 minutes is not altogether unexpected since it is likely to form a uniform 3D structure of cement particles flocculated by bridging SAA molecules. It is interesting to note that the latex particles decrease the elasticity present in the OS sample. This is probably due to the polydispersity in the dispersion and the reduced number of elastic bonds between particles.

The oscillation experiments give a measure of the potential of the system to store energy (G', the storage modulus) and to dissipate energy (G'', the loss modulus). Thus it may be inferred that the former will be related to the floc-floc interactions in the pastes, while the latter will be related to flow and hence workability. From Figs. 3(b) and 4(b) it can be seen that the control paste very rapidly forms a flocculated network of particles. However, it appears from the stress relaxation data in Fig. 2(a) that the interactions between the particles are relatively weak. The rapid increase at about 80 minutes may indicate the onset of secondary hydration processes not detectable in the stress relaxation experiments. Another interesting feature of the storage modulus data involves the samples containing high levels of SAA (ie. 10/1/10 and 10/2/10). The relatively low values of G' probably relates to the complete dispersion of both the latex and cement particles, and screening of the interactions between the particles by adsorbed SAA molecules.

From Figs. 3(c) and 4(c) it is evident that the control paste becomes more fluid (ie. smaller G'' value) at the higher frequency. Also it is apparent that samples 10/1/5 and 10/2/5 become much less fluid at the higher frequency. The control becomes more fluid due to the increased energy input at the higher frequency which will cause the flocculated structure to break up and occluded water to be released. The samples containing low levels of surfactant became less fluid presumably because the increased energy input enhances the flocculation mechanism involving the adsorbed SAA molecules. The latter decrease in the loss modulus at the high frequency (eg. 10/1/5) is probably due to slippage in the rheometer. The low values of G'' for the samples containing excess SAA indicate that the samples remain fluid over the time of study. However, it can be seen that at any given time the value of G'' is greater at the higher frequency for sample 10/1/10.

This paper illustrates the use of different rheological techniques to yield information on both the fundamental interactions in cement pastes and the bulk properties of the systems. The data confirms the complex nature of the effect of the addition of latex particles (+SAA) to cement pastes.

Acknowledgement

The authors would like to thank the Science and Engineering Research Council, UK for financial support.

References

Banfill, P.F.G. (1981) A Viscometric Study of Cement Pastes Containing Superplasticisers With a Note on Experimental Techniques. **Mag. Concr. Res.**, 33, 37-47.

Banfill, P.F.G. and Gill, S.M. (1986) The Rheology of Aluminous Cement Pastes. **Proc. 8th Int. Congr. Chem. Cem.**, Rio de Janeiro, Brazil.

Banfill, P.F.G. and Saunders, D.C. (1981) On the Viscometric Examination of Cement Pastes. **Cem. Concr. Res.**, 3, 363-370.

Bhatty, J.I. and Banfill, P.F.G. (1984) A Viscometric Method of Monitoring the Effect of Retarding Admixtures on the Setting of Cement Pastes. **Cem. Concr. Res.**, 14, 64-72.

Ferry, J.D. (1980) Viscoeleastic Properties of Polymers. Wiley, N.Y.

Gregory, T. and O'Keefe, S.J. (1988) The Rheological Properties of Fresh Polymer-modified Cement Pastes. **Proc. 5th Int. Congr. Rheol.**, Sydney, Australia.

Hannant, D.J. and Keating, J. (1985) Equipment for Assessing the Development of Structure in Fresh Cement Pastes by the Measurement of Shear Modulus. **Cem. Concr. Res.**, 15, 605-612.

Lapasin, R., Papo, A. and Rajgelj, S. (1983) Flow Behaviour of Fresh Cement Pastes. A Comparison of Different Rheological Instruments and Techniques. **Cem. Concr. Res.**, 13, 349-356.

PART TWO
THEORETICAL STUDIES

10 A NEW VISCOSITY EQUATION FOR NON-NEWTONIAN SUSPENSIONS AND ITS APPLICATION

K. HATTORI
Department of Civil Engineering, Chuou University,
Tokyo, Japan
K. IZUMI
Patent Department, Kao Corporation, Tokyo, Japan

Japan

Abstract
A new rheological theory was developed by introducing the concept of friction between components of a suspension into classical Newton's flow equation. By this new theory, the flow properties of non-Newtonian suspensions became explainable on the same basis as Newtonian fluids.
By combining effects of the natural coagulation and of the mechanical dissociation of particles in a suspension, the equation which was capable of describing the time-dependent and shear-dependent viscosities was obtained. By using this eqation, the viscosity of suspension at any given time t became calculable from the known values of parameters B_3, H, n_3, U_0, and γ.
It became also possible by the present theory to calculate flow curves and hysteresis loops by introducing a shearing condition into the general viscosity equation and using it in the Newtons flow equation (1).

$$\tau = \eta \cdot \gamma \qquad (1)$$

Flow types of suspensions conventionally used to classify the fluids such as dilatant and thixotropic were found to vary from one to another under the influences of experimental conditions, and the variation of flow types under this influences became predictable by calculation in the present theory.
Keywords: Non-Newtonian, Suspension, Coagulation rate theory, Structure of coagulation, Junction between particles, Flow curve, Flow type, Dilatancy, Thixotropy, Hysteresis, Time-dependent viscosity, Viscosity equation.

1 Introduction

By introducing the concept of friction between molecules[1,2], the viscosity (η) of a Newtonian liquid is expressed as follows;

$$\eta = B \cdot n^{2/3} \qquad (2)$$

B : coefficient of friction between liquid molecules
n : number of liquid molecules in unit volume

By applying the same concept to the friction points or the junctions between components of a suspension and by defining the parameter U_3 which is a dimensionless variable corresponding to the ratio J_t/n_3 (J_t

is the number of junctions between particles at any given time t and is explained further in the later part of this paper), eq.(2) is extended to express viscosities of suspensions (η_s) as follows;

$$\eta_s = \eta_1 + \eta_2 + \eta_3$$
$$= B_1 \cdot n_1^{2/3} + B_2 \cdot n_2^{2/3} + B_3 \cdot (n_3 \cdot U_3)^{2/3} \tag{3}$$

Suffixes $_1$, $_2$, $_3$ stand for attributes of parameters, $_1$ for liquid molecules, $_2$ for liquid molecules and particles, and $_3$ for particles in suspension respectively.

For the uniformity of expression, dimensionless constants U_1 and U_2 (both are equal to 1) may be introduced into eq. (3) and Newtonian flow equation (1) is extended to non-Newtonian suspensions as follows;

$$\tau = \gamma \cdot \{B_1 \cdot (n_1 \cdot U_1)^{2/3} + B_2 \cdot (n_2 \cdot U_2)^{2/3} + B_3 \cdot (n_3 \cdot U_3)^{2/3}\} \tag{4}$$

Since U_3 is the only variable in eqs. (3) and (4), the viscosities and the flow curves of suspensins become calculable if a practical form of equation expressing U_3 is obtained.

2 An equation for the calculation of increasing junction number

As the basis for deriving equations expressing the varying number of junctions, the theory of peri-kinetic coagulation rate established by Verwey and Overbeek[3] was employed. By their theory, the decreasing number of particles in suspension the particles in which are completely dispersed ($n_t = n_3$) initially, is expressed as follows;

$$-\frac{dn_t}{dt} = H \cdot n_t^2 / n_3 \tag{5}$$

n_3 : primary particle number
n_t : particle number at a given time t
H : coagulation rate constant of particles

By integrating eq. (5), we have,

$$n_t = \frac{n_3}{Ht + 1} \tag{6}$$

Since eqs.(5) and (6) were obtained on the assumption[3,4,5] that a junction is formed by a collision and that the resulting structure of coagulated particles is a chain, the following relationship between the particle number and the junction number is maintained during the process of coagulation,

$$J_t + n_t = n_3 \tag{7}$$

and from eq.(7), we have the equation for the calculation of junction number at any given time t.

$$J_t = \frac{n_3 \cdot Ht}{Ht + 1} \qquad (8)$$

3 An equation expressing influence of the shear rate

From dimensional analysys of the coagulation rate constant H, it is known that the rate of coagulation or the increasing number of junction is governed by the ratio of energies which may be called a dimensionless energy. From the dimensional analysis of eq. (1), it is known that the physical dimension of shear rate is the same as that of H the dimension of which is $[T^{-1}]$, containing the expression of dimensionless energy as shown by the following equation.

$$\gamma = \frac{\tau \, (dyn/cm^2)}{\eta \, (dyn \cdot sec / cm^2)} = \frac{\tau \, (dyn \cdot cm/cm^3)}{\eta \, (dyn \cdot cm \cdot sec / cm^3)} = \frac{\tau \, (erg)}{\eta \, (erg \cdot sec)} \qquad (9)$$

The same expression is obtainable by denoting the total mechanical energy (of agitation) E_m put into the system (of viscosity measurement) during the period of time t and dividing it by kT to be the same as the expressions for energy in the coagulation rate theory.

$$\gamma = \frac{E_m}{kTt} \quad \text{or} \quad \gamma t = \frac{E_m}{kT} \qquad (10)$$

As the state of dispersion is described by the relationship between the potential energy and the distance between particles, it may be considered that the coagulation progresses by consuming the internal energy corresponding to H·t as expressed by eq. (8). To satisfy the law of conservation of energy and to maintain the relationship between the internal energy and the state of dispersion, the original state before consuming the internal energy corresponding to H·t must be recovered if the same amount of energy $\gamma \cdot t'(= H \cdot t)$ is supplied to the system from outside. For this reason, the expression for the influence of energy of agitation on the particle number is obtainable by back-tracing the curve expressed by eq. (8).

For back-tracing the curve expressed by eq. (8), the relationship between the internal energy consumption and the external energy to recover the original state (H·t = $\gamma \cdot t'$) must be modified as follows;

$$H \cdot t = 1/\gamma \cdot t \quad (H = 1/\gamma \cdot t^2) \qquad (11)$$

By reversing the time axis and by introducing the relationship of eq. (11) into eq. (5), we have,

$$\frac{dn_t}{dt} = \frac{n_t^2}{n_3 \cdot \gamma \cdot t^2} \qquad (12)$$

which integrates to,

$$n_t = \frac{n_3 \cdot \gamma \cdot t}{\gamma \cdot t + 1} \qquad (13)$$

for the initial conditions $t = \infty$ and $n_t = n_3$. From eq. (13), we have the following (14) for expressing the influence of agitation in promoting the dispersion.

$$J_t = \frac{n_3}{\gamma \cdot t + 1} \quad \text{or} \quad \frac{J_t}{n_3} = U_3 = \frac{1}{\gamma \cdot t + 1} \tag{14}$$

Although eq. (14) expresses the rate of mechanical dissociation of the coagulated particles, it is not useful for the calculation of decreasing viscosity of actual suspension because the parameter γ only desacribes the mechanical dissociation rate. In the suspensions under the influence of the coagulation rate H, the natural coagulation of particles always progresses until the complete coagulation is reached, so, it is always necessary to have the parameter H in the equations to calculate the time-dependent viscosity.

4 Viscosity equations for suspensions

For the suspensions, two initial states, the initially dispersed state and the initially coagulated state may be defined. For an initially dispersed suspension left unagitated, the coagulation progresses and the junction number increases according to eq. (8). If all the particles in suspension are completely dispersed, there is no junction to be destroyed by the agitation, but considering the process of preparing a suspension, the probability of finding or observing the completely dispersed state seems very low.

Although the states of complete dispersion or complete coagulation is not practically observable, we may define the theoretical time axsis in which the suspensions initially have to be either in the state of complete coagulation in which J_t is equal to n_3 and n_t is equal to 0 or in the state of complete dispersion ($J_t = 0$, $n_t = n_3$).

For the chain structure of coagulated particles, the following relationship between the particle number and the junction number is always satisfied.

$$J_0 + n_0 = n_3, \qquad U_0 = J_0 / n_3$$
$$J_t + n_t = n_3, \qquad U_3 = J_t / n_3$$
$$J_E + n_E = n_3, \qquad U_E = J_E / n_3$$

J_0 ; Junction number at the start of observation.
J_t ; Junction number at the experimental time t.
J_E ; Junction number at the time of equilibrium (explained later)
n_0 ; Particle number at the start of observation.
n_t ; Particle number at the experimental time t.
n_E ; Particle number at the time of equilibrium (explained later)

Since the initial conditions of suspensions are not usually observable, it is important to note that almost all experimental data are described or illustrated on the experimental time axsis. The following FIG. 1 illustrates the difference between the theoretical time and the experimental time in which the observable viscosity changes are shown by solid lines.

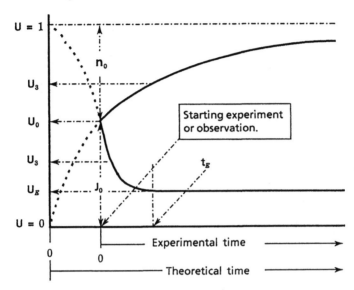

FIGURE 1. DEFINITION of TIME AXSIS

For the n_0 particles, the coagulation progresses at the rate of $Ht/(Ht+1)$, but by the agitation (of rotor of the viscometer used for the measurement), the rate is reduced by the ratio corresponding to $1/(\gamma t+1)$. Consequently, the number of junction J_1 for n_0 particles at the time t is;

$$J_1 = \frac{n_0 \cdot Ht}{(Ht + 1)(\gamma t + 1)} \tag{15}$$

For the J_0 junctions, the dissociation progresses at the rate $1/(\gamma t+1)$, and for the released particles by the dissociation $\{1 - 1/(\gamma t+1)\}$, the coagulation progresses at the rate $Ht/(Ht+1)$. So the junction number J_2 junctions at the time t originating from J_0 is,

$$J_2 = \frac{J_0}{\gamma t + 1} + \frac{J_0 \cdot \gamma Ht^2}{(Ht + 1)(\gamma t + 1)} = J_0 \left\{ 1 - \frac{\gamma t}{(Ht + 1)(\gamma t + 1)} \right\} \tag{16}$$

Since the junction number at the time t is the sum of J_1 and J_2,

$$J_t = J_0 \left\{ 1 - \frac{\gamma t}{(Ht + 1)(\gamma t + 1)} \right\} + \frac{n_0 \cdot Ht}{(Ht + 1)(\gamma t + 1)}$$

$$= \frac{J_0 \cdot (\gamma Ht^2 + 1) + n_3 \cdot Ht}{(Ht + 1)(\gamma t + 1)} = n_3 \cdot \frac{U_0 \cdot (\gamma Ht^2 + 1) + Ht}{(Ht + 1)(\gamma t + 1)} \tag{17}$$

By substituting the initial conditions ($U_0 = 1$) or ($U_0 = 0$) into eq. (17), equations for the initially coagulated suspension and the initially dispersed suspension previously reported[1,2] are obtainable.

As the coagulation progresses, the number of isolated particles diminishes reducing the probability of collision between particles. As the dissociation of coagulated particles progresses, the rate of coagulation increases due to the increased number of particles and to the increased probability of collision. For this reason, there exists a time of equilibrium t_E at which the rates of coagulation and of mechanical dissociation become equal. After the time of equilibrium, the number of junctions remains constant. The time of equilibrium is calculable from the condition to make the derivative of eq. (17) equal to zero.

The derivative of eq. (17) is,

$$\frac{dJ_t}{dt} = n_3 \cdot \left[\frac{(\gamma Ht^2 - 1)\{U_0 \cdot (\gamma + H) - H\}}{(Ht + 1)^2 (\gamma t + 1)^2} \right] \tag{18}$$

and the time of equilibrium is,

$$t_E^2 = \frac{1}{\gamma \cdot H} \qquad \text{or} \qquad t_E = \frac{1}{\sqrt{\gamma \cdot H}} \tag{19}$$

The initial increase or the decrease of viscosity by the general equation (17) depends on both (γHt^2) and $H/(\gamma + H)$. When U_0 is larger than $H/(\gamma + H)$. the viscosity initially decreases since (γHt^2) is usually very small when t is nearly 0. The viscosity in this case decreases until the the time of equilibrium and remains constant afterwards. If U_0 is smaller than $H/(\gamma + H)$. the viscosity increases until the equilibrium is reached.

The number of junctions J_E at the time of equilibrium is calculable by substituting t_E into eq. (17).

$$J_E = n_3 \cdot \left\{ \frac{2 \cdot U_0 \cdot \sqrt{\gamma \cdot H} + H}{(\sqrt{H} + \sqrt{\gamma})^2} \right\} \tag{20}$$

After obtaining the equations for the calculation of junction numbers J_t and J_E, the viscosity of a suspension originating from the particle-particle friction becomes calculable as follows based on eqs. (2) and (3).

$$\left. \begin{array}{l} \eta_3 = B_3 \cdot J_t^{2/3} \\ \eta_E = B_3 \cdot J_E^{2/3} \end{array} \right\} \tag{21}$$

5 Calculation of flow curves and hysteresis loops

After obtaining the general viscosity equation (17), flow curves and hysteresis loops become calculable by the extended Newton's flow equation (4). Before the calculation, the experimental (shearing) condition has to be defined. Although there is no restriction in selecting the shearing condition, a complicated condition is usually avoided and a simple conditions such as the linear change of shear rate expressed as follows are frequently adopted.

$$\gamma = G \cdot t \tag{22}$$
$$\gamma = G \cdot (T - t), \qquad T \geq t \geq T/2 \tag{23}$$

G : dimensionless constant determining the shear changing rate
T : total cycle time of hysteresis experiment

By introducing shearing conditions (22) and (23) into general viscosity equation (17), we have,

$$\eta_s = B_s n_s^{2/3} \cdot \left\{ \frac{U_a \cdot (GHt^3 + 1) + Ht}{(Ht + 1)(Gt^2 + 1)} \right\}^{2/3} \tag{24}$$

$$\eta_s = B_s n_s^{2/3} \cdot \left\{ \frac{U_b \cdot (GHt^2 \cdot (T - t) + 1) + Ht}{(Ht + 1) \cdot \{G \cdot (T - t) \cdot t + 1\}} \right\}^{2/3} \tag{25}$$

U_a, U_b : Correspond to U_0 in FIG. 1. Used to discriminate initial conditions for increasing shear rate (up-curve) and decreasing shear (down-curve) rate. U_b is calculable from final viscosity on the up-curve.

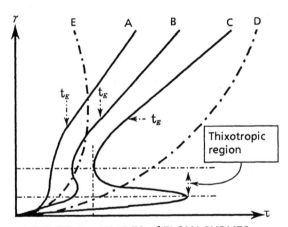

FIGURE 2. SHAPES of FLOW CURVES

By using eqs. (22) and (24), flow curves for the increasing shear rate (A, B, and C in FIG. 2) are calculable ($\tau = \eta_3 \cdot \gamma$). In the same manner, the curves for the decreasing shear rate (D and E in FIG. 2) are calculable by eqs. (23) and (25). By connecting an up-curve and a down-curve at T/2, a hysteresis loop is obtainable as shown below in FIG. 3.

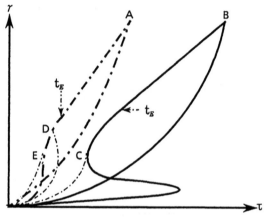

FIGURE 3. HYSTERESIS LOOPS

As shown in FIG. 3, the shapes of hysteresis loops vary depending on the total cycle time and the turning point of the experiments where the increasing shear rate turns to decrease which are independent of the quality of sample suspensions. Particularly, the shape changes largely depending on whether the turning points (A, B, C, D, and E in FIG. 3) come before the time of equilibrium t_E or after.

6 Analysis of flow curves

Although flow curves and hysteresis loops are calculable as shown in FIGs. 2 and 3 by the equations in the present theory, it seems better to use extended Newton's flow equation in differential form (26) for analyzing the influences of parameters H, U_0, γ, and G on the shapes of flow curves.

$$\frac{d\tau}{dt} = \frac{d(\eta_s \cdot \gamma)}{dt} = \eta_s \frac{d\gamma}{dt} + \gamma \frac{d\eta}{dt}$$

$$= (\eta_1 + \eta_2 + \eta_3) \cdot \frac{d\gamma}{dt} + \gamma \frac{d\eta_3}{dt} \tag{26}$$

Since the shapes of flow curves calculated by eq. (24) in the time-shear stress plane is the same as that in the shear rate-shear stress plane when the shear rate inceases proportionally to the time, eq. (26) is useful for predicting the shapes of flow curves, and

ince n_1 and n_2 are not influential to the shapes, these parameters
ay be omitted from the differential equation expressing $d\tau/dt$.

$$\gamma = G \cdot t \qquad d\gamma/dt = G$$

$$\frac{dn_3}{dt} = \frac{2 \cdot B_3 \cdot n_3^{2/3}}{3} \cdot \left\{ \frac{(G \cdot t^2 + 1) \cdot (Ht + 1)}{U_a(GHt^3 + 1) + Ht} \right\}^{1/3}$$

$$\cdot \frac{G \cdot t \cdot (U_a GHt^3 - Ht - 2U_a) + H \cdot (2GHt^3 - 1) \cdot (U_a - 1)}{(G \cdot t^2 + 1)^2 \cdot (Ht + 1)^2}$$

$$(27)$$

$$\gamma = G \cdot (T - t), \qquad d\gamma/dt = -G$$

$$\frac{dn_3}{dt} = \frac{2 \cdot B_3 \cdot n_3^{2/3}}{3} \cdot \left[\frac{(Ht + 1) \cdot \{G \cdot (T - t) \cdot t + 1\}}{U_b \cdot (GHt^2 \cdot (T - t) + 1) + Ht} \right]^{1/3} \cdot$$

$$\frac{G \cdot [GHt^2 U_b (T-t)^2 + (T-2 \cdot t) \cdot \{U_b (H^2 t^2 - 1) - H^2 t^2\} + Ht^2] + (1 - U_b)H}{(Ht + 1)^2 \cdot \{Gt \cdot (T - t)^2 + 1\}^2}$$

$$(28)$$

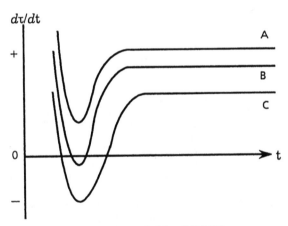

FIGURE 4. $d\tau/dt$ CURVES

FIG. 4 shows typical $d\tau/dt$ curves calculated by eq.(27). From the
experience of calculation of flow curves using eq. (4), it was known
that the variation in the shapes of flow curves was caused by the
changing values of parameters H, G, U_0 and T. It is also an
experimental experience that H value for a variety of aqueous
suspensions were usually in the range between about 10^{-2} and 10^{-9}.
   When H value is very large, $d\tau/dt$ remains to be positive as shown
by the curve A in FIG. 4. As H decreases, the minimum value of $d\tau/dt$
decreases and the curve passes through the negative region as shown

by curves B and C in FIG. 4. The variation of $d\tau/dt$ from A to C corresponds to the change of flow curves A to C shown in FIG. 2.

Contrary to the variation of H, the $d\tau/dt$ curve passes through the negative region when G value is large enough. As G decreases, the $d\tau/dt$ value increases and the thixotropic region on the flow curve shown in FIG. 2 disappears.

The variation of $d\tau/dt$ value caused by the changing value of U_0 is directionally similar to that caused by the variation of G. When U_0 is close to 0 (U_0 takes the value between 0 and 1) the $d\tau/dt$ value remains to be positive, and when it is close to 1, the $d\tau/dt$ curve passes through the negative region.

The influence of T(or T/2) was already explained by FIG. 3.

The difference between the down curves D and E in FIG. 2 is also predictable by calculation using eq. (28) in the same manner as described above.

7 Conclusion

Decreasing H value causes variations in the shapes of flow curves from C to A as shown in FIG. 2. H value tends to decrease by the addition of effective dispersant, by the lower concentration of particles, and by the higher viscosity of medium. It tends to increase by the higher salt concentration. There may be other factors which cause the change of H value, but these may not be adjustable artificially.

Increase in G value causes variations in the shapes of flow curves from A to C. Larger G value is achieved by setting rapid change of the shear rate.

Increase in U_0 value causes variations in the shapes of flow curves from A to C. U_0 value increases when the suspension sample is left unagitated before the rheological experiment. The longer the unagitated time, the closer the U_0 value to 1.

Among the parameters causing variarions in the shapes of flow curves of suspensions, H is only one which is related to the chemical conposition. Since other parameters are not related to the composition and are simple experimental conditions, it is important to note that the variation in the shapes of flow curves does not always describe the difference in the quality of suspension samples.

8 References
1)K. Hattori, K. Izumi, (1982) J. Dispersion Sci. Tech., 3, (2), 129
2)K. Hattori, K. Izumi, (1982) "Proc. Symposium M", Annual Meeting
   of Materials Res. Soc., Edited by J. P. Skalney : Nov. 1-4,
   Boston Mass. U. S. A., pp. 14.
3)E. J. W. Verwey, J. Th. G. Overbeek, (1948), "Theory of the
   Stability of Lyophobic Colloids", Elsvier Publishing Co., Ltd.
4)M. von Smoluchowski, (1916), Phys. Z., 17, 557, 585
5)N. Casson, (1959), "Rheology of Disperse Systems", ed. by C.C.Mill
   Pergamon Press, London, pp. 84

11 METHOD OF DETERMINING RHEOLOGICAL PARAMETERS FOR LINEAR VISCOELASTIC MEDIUM UNDER FOUR KINDS OF UNIFORM LOADING

A. TERENTYEV and G. KUNNOS
Technical University of Riga, Latvia, USSR

Abstract
Interrelationships for elastic, viscous parameters and Poisson's ratio analog of linear isotropic viscoelastic media under four types of uniform loading - shear, three-dimensional, one-dimensional, uniaxial - have been established using elastoviscous operational analogy. The established relationships are of highly practical value as it is enough to know two of the six sets of values, i.e. to carry out two tests (or even one) to determine the whole group. For the practical implementation of the method to obtain state equations of the material, its rheological model and the complete set of their respective parameters experiments upon swelling gas concrete mix were made. Valid data have been obtained by test and calculation for six sets of rheological parameters of the mix and their variations in time. Simple shear and one-dimensional loading were selected as the initial types of loads.
Key words: Linear isotropic viscoelastic medium, Uniform loading, Rheological parameters, Gas concrete mix.

1 Introduction

Generalized Hooke's law for an isotropic elastic body contains two independent constants characterizing changes in shape and volume of the material under external load. In order to determine these constants two independent experiments are to be made for material strain under a load where changes take place both in shape and volume. In this case the simplest and most unambiguous tests are those for uniform loading where strain is proportional to the stress applied and equal throughout the bulk of the material. Four types of uniform loading of the material are known. Volume (spherical) and shear (deviatoric) loadings are the principal pure types, and any kind of combined load can be expressed in terms of the two types. The other two, namely, one-dimensional (compressive) and uniaxial loads are of combined character where strain includes changes in both volume and shape.

Every type of material strain is characterized by its specific set of rheological parameters. The methods for determining the rheological parameters in visco-elastic media are known for all four principal types of uniform loads. Correspondingly, there are four independent tests to be made followed by creep analysis of the test curves for the material tested. In the result the whole set of rheological properties of the material is obtained.

The above methods require four independent experiments to be made using four different devices which is a greatly labour-consuming procedure, especially if the experiments are to be carried out simultaneously due to the time-dependent character of the medium, e.g. in the hardening of material. Besides it is difficult and sometimes impossible to apply certain types of uniform load to certain materials, like uniaxial load to liquid media, or universal compression to expanding media, like gas concrete mix.

In the elasticity theory definite relationships are known to exist between elasticity moduli for every type of loading and Poisson's ratio. These relationships enable us to calculate the three unknown values if any two other values are given (for example, see the relationship Table 1 | 1,2 |).

Table 1. Relationship between elastic moduli and Poisson's ratio

GIVEN VALUES	G	K	M	E	E_v	V
1. G,K	-	-	$K + \frac{4}{3}G$	$\frac{9KG}{3K+G}$	$\frac{18KG}{3K-2G}$	$\frac{3K-2G}{2(3K+G)}$
2. G,M	-	$M - \frac{4}{3}G$	-	$\frac{G(3M-4G)}{M-G}$	$\frac{2G(3M-4G)}{M-2G}$	$\frac{M-2G}{2(M-G)}$
3. G,E	-	$\frac{EG}{3(3G-E)}$	$\frac{G(4G-E)}{3G-E}$	-	$\frac{2EG}{E-2G}$	$\frac{E-2G}{2G}$
4. G,E	-	$\frac{2E_vG}{3(E_v-6G)}$	$\frac{2G(4G-E_v)}{6G-E_v}$	$\frac{2E_vG}{E_v-2G}$	-	$\frac{2G}{E_v-2G}$
5. G,V	-	$\frac{2G(1+V)}{3(1-2V)}$	$\frac{2G(1-V)}{1-2V}$	$2G(1+V)$	$\frac{2G(1+V)}{V}$	-
6. K,M	$\frac{3}{4}(M-K)$	-	-	$\frac{9K(M-K)}{M+3K}$	$\frac{9K(M-K)}{3K-M}$	$\frac{3K-M}{3K+M}$
7. K,E	$\frac{3KE}{9K-E}$	-	$\frac{3K(3K+E)}{9K-E}$	-	$\frac{6EK}{3K-E}$	$\frac{3K-E}{6K}$
8. K,E_v	$\frac{3KE_v}{2(9K+E_v)}$	-	$\frac{3K(3K+E_v)}{9K+E_v}$	$\frac{3KE_v}{6K+E_v}$	-	$\frac{3K}{6K+E_v}$
9. K,V	$\frac{3K(1-2V)}{2(1+V)}$	-	$\frac{3K(1-V)}{1+V}$	$3K(1-2V)$	-	-
10. M,E	$\frac{3M+E-W}{8}$	$\frac{3M-E-W}{6}$	-	-	$\frac{EM}{E-M-W}$	$\frac{E-M+W}{4M}$
11. M,V	$\frac{(1-2V)M}{2(1-V)}$	$\frac{(1+V)M}{3(1-V)}$	-	$\frac{(1-2V)(1+V)M}{1-V}$	$\frac{(1-2V)(1+V)M}{(1-V)V}$	-
12. E,E_v	$\frac{EE_v}{2(E_v+2E)}$	$\frac{3E_vE}{9(E_v-2E)}$	-	-	-	$\frac{E}{E_v}$
13. E,V	$\frac{E}{2(1+V)}$	$\frac{E}{3(1-2V)}$	$\frac{(1-V)E}{(1+V)(1-2V)}$	-	$\frac{E}{V}$	-
14. E_v,V	$\frac{E_vV}{2(1+V)}$	$\frac{E_vV}{3(1-2V)}$	$\frac{(1-V)VE_v}{(1+V)(1-2V)}$	E_vV	$W = \sqrt{(M-E)(9M-E)}$	

A similar table for linear isotropic media would be of greatest value in that it would permit us to obtain the full set of elastic and viscous properties for the four types of loads from only one or two experiments for uniform loading of the material; the theoretical supplements would speak for themselves if such existed.

Therefore our purpose was to find the relationship between elastic and viscous characteristics in linear isotropic media under the four types of uniform loading. Expression of elastic and viscous constants referring to one type of loading by elastic and viscous constants referring to the other two types of load is not trivial, however, because the relationships sought for viscoelastic media should be in fact relationships between mathematical structures containing elastic and viscous constants, their number being arbitrary (whereas in elasticity theory their number is finite).

2 Elastoviscous analogy

Let us describe viscoelastic medium in the most general way adopting an operator form of expressing the relations between stresses and strains for deviatoric and spherical loading:

$$\bar{P}\, S_{ij} = 2\, G_\infty \bar{Q}\, e_{ij} \qquad (1)$$

$$\bar{R}\, \mathfrak{S}_{kk} = 3\, K_\infty \bar{T} \varepsilon_{kk} \qquad (2)$$

where S_{ij} and $\mathfrak{S}_{kk}$ are deviatoric and spherical stress tensors, e_{ij} and δ_{kk} are deviatoric and spherical strain tensors, G_∞ and K_∞ are continuous elasticity moduli for simple shear and universal loading; $\bar{P}$, $\bar{Q}$, $\bar{R}$ and $\bar{T}$ are some linear operators .

It should be noted here that proceeding from test exemplification for spherical loading the relationship (2) in a general form practically is never written. Introducing the following notations for linear fractional operators

$$\hat{G} = G_\infty \, \frac{\bar{Q}}{\bar{P}} \; ; \quad \hat{K} = K_\infty \, \frac{\bar{T}}{\bar{R}} \qquad\qquad (3)$$

generalized Hooke's law for linear isotropic viscoelastic media will be given in the form of generalized Hooke's law for elastic isotropic media:

$$\mathfrak{S}_{ij} = \hat{\lambda} \, \delta_{ij} \, \mathfrak{e}_{kk} + 2 \, \hat{G} \mathfrak{e}_{ij} \qquad\qquad (4)$$

where $\hat{G}$ and $\hat{\lambda} = \hat{K} - 2/3 \, \hat{G}$ respectively are analogs of Lamee coefficients in operational form. Further it is completely evident that the familiar relationships of elasticity theory in operational terms will be valid also for linear viscoelastic media.

Thus, specifying the principle of elastoviscous analogy | 3 | in the following expressions:

$$\hat{M} = M_\infty \, \frac{\bar{D}}{\bar{C}} \; ; \quad \hat{E} = E_\infty \, \frac{\bar{B}}{\bar{A}} \; ; \quad \hat{E}_\nu = E_{\nu\infty} \, \frac{\bar{H}}{\bar{F}} \; ; \quad \hat{\hat{\nu}} = \frac{\hat{E}}{\hat{E}_\nu} \; ; \quad (5)$$

and considering also (3), where

$\hat{M}$ is operator of one-dimensional loading,

$\hat{E}$ is operator of longitudinal uniaxial loading,

$\hat{E}_\nu$ is operator of transversal uniaxial loading,

$\hat{\hat{\nu}}$ is operator of transversal strain,

we obtain the interrelation between bilinear operators corresponding to the four types of uniform loading of a linear viscoelastic material where they are similar in form with the relationship between elasticity moduli and Poisson's ratio in elasticity theory, i.e., Table 1 can be used for a linear viscoelastic body where elasticity coefficients are replaced by the corresponding operators from expressions (3) and (5). For instance, in (4) $\hat{\lambda}$ and $\hat{G}$ can be found from the table of viscoelastic operator relationships considering the two operators known, and it means obtaining a specific form of generalized Hooke's law for the given linear viscoelastic body described by two given initial operators.

3 Specification of operational form

Let us represent linear operators $\bar{A}, \bar{B}, \bar{C}, \bar{D}, \bar{E}, \bar{H}, \bar{P}, \bar{Q}, \bar{R}, \bar{T}$ in the following differential form:

$$\bar{A} = a_o + a_1 \frac{d}{dt} + a_2 \frac{d^2}{dt^2} + \ldots + a_n \frac{d^n}{dt^n} \; ,$$

$$\bar{B} = b_o + b_1 \frac{d}{dt} + b_2 \frac{d^2}{dt^2} + \ldots + b_m \frac{d^m}{dt^m} \; ,$$

$$\ldots\ldots\ldots\ldots\ldots\ldots\ldots\ldots\ldots\ldots\ldots\ldots\ldots\ldots \quad (6)$$

$$\bar{T} = t_o + t_1 \frac{d}{dt} + t_2 \frac{d^2}{dt^2} + \ldots + t_p \frac{d^p}{dt^p} \; ,$$

where $a_o, a_1, a_2, \ldots, a_n$

$b_o, b_1, b_2, \ldots, b_m$ $\qquad\qquad\qquad\qquad (7)$

$\ldots\ldots\ldots\ldots\ldots$

$t_o, t_1, t_2, \ldots, t_p$ are rheological parameters.

This enables us to describe rheological behaviour of media by means of visual mechanical models corresponding to each bilinear operator by using conventional viscosity and elasticity parameters, unlike the integral approach.

4 Practical implementation of the method

Let us discuss the method of obtaining state equations of the material, its rheological models and their respective parameters under the four types of uniform loading; we shall proceed from the data known for the two loading types Let us make a practical implementation of the method by help of test data. The material to be investigated was swelling gas concrete mix with time-depende properties. The implementation procedure is the following.
Out of the four types of tests only two are carried out, e.g., one-dimensional and shear loading.
In the first experiment the mix to be investigated is placed in the odometer cylinder and a balanced piston placed onto its surface; the piston is connected to an external device. The mix is subjected to one-dimensional load $G(t)$ using a standard method | 4 |. Periodical instantaneous loading mode is applie by permanent load G , the exposure lasts for some time, and the mix is unloaded instantaneously. The piston movements and the rise in temperature are recorded. Fig. 1 shows a part of the strain curve obtained.

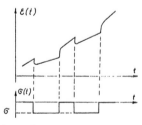

Fig.1. One-dimensional loading

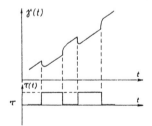

Fig.2. Shear

1 the second experiment the same mix composition is placed into the cuvette
f Weiler-Rebinder apparatus. The cuvette is placed into the vessel of the
racing liquid thermostat which simulates the temperature regime registered in
he first experiment with the preset degree of accuracy. Shear loading of the
ix is applied by standard methods | 5 | using a balanced knurled plate. The
oading mode τ (t) is periodical instantaneous loading by permanent shear
oad τ , exposing and unloading the mix instantanously. The plate movements
re recorded. Fig. 2 shows a part of the strain curve.
fter that an adequate rheological model is selected for every creep test curve
Figs. 1, 2).
rheological model of a standard linear viscoelastic body shown in Fig. 3
orresponds to creep curve δ (t) in Fig. 1 where the swelling is not
onsidered, and is described by equation

$$G + n_1^M \dot{G} = M_\infty \delta + M_\infty n_2^M \dot{\delta} \qquad (8)$$

then

$$\hat{M} = M_\infty (1 + n_2^M \frac{d}{dt}) / (1 + n_1^M \frac{d}{dt}) \qquad (9)$$

here $n_2^M = \eta^M / M_2$ is retardation time and $n_1^M = n_2^M M_\infty / M_1$ is relaxation time
he values of the given coefficients are calculated from the test curve in Fig.
sing the conventional methods | 6 |.
he creep test curve γ (t) where swelling is again not considered correspond
o the rheological model of a standard linear viscoelastic body as seen in Fig.
nd is described by equation

$$\tau + n_1^G \dot{\tau} = G_\infty \gamma + G_\infty n_2^G \dot{\gamma} \qquad (10)$$

then

$$\hat{G} = G_\infty (1 + n_2^G \frac{d}{dt}) / (1 + n_1^G \frac{d}{dt}) \qquad (11)$$

here $n_2^G = \eta^G/G_2$ is retardation time and $n_1^G = n_2^G G_\infty / G_1$ is relaxation time.
he values of the above expressions are determined from the test curve of Fig. 2
sing the conventional methods | 6 |.
n this way operators $\hat{M}$ and $\hat{G}$ have been found. Further, formulas from Table 1
ive us a clue for finding operators $\hat{K}$, $\hat{E}$, $\hat{E}_\nu$, $\hat{\nu}$ i.e., operators $\bar{R}$, $\bar{T}$, $\bar{A}$, $\bar{B}$, $\bar{F}$,
 and their coefficients r_i, t_i, a_i, b_i, f_i and h_i, which themselves are in
act the rheological parameters.

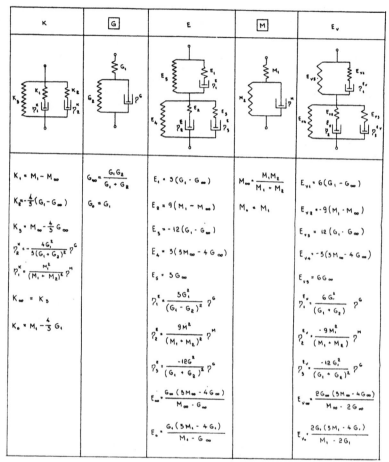

Fig. 3. Correspondence between rheological models for linear isotropic viscoelastic medium under four types of uniform loading.

5 Examples

The stess-strain relations for three-dimensional loading are written in opera form like

$$P = 3 \hat{K} \Theta, \qquad (12)$$

where Θ is mean volume strain, and P is volume load.
By substituting the corresponding operators from Table 1 for elastic moduli we obtain the following equation of state for three-dimensional loading with elastoviscous parameters determined experimentally for one-dimensional and shear loading:

$$P + \left(n_1^M + n_1^G \right) \frac{dP}{dt} + n_1^M n_1^G \frac{d^2P}{dt^2} = 3 \left[\left(M_\infty - \frac{4}{3} G_\infty \right) \Theta + \right.$$

$$\left[\left(M_1 - \frac{4}{3} G_\infty \right) n_1^M + \left(M_\infty - \frac{4}{3} G_1 \right) n_1^G \right] \frac{d\Theta}{dt} + \left(M_1 - \frac{4}{3} G_1 \right) n_1^M n_1^G \frac{d^2\Theta}{dt^2} \right] \qquad (13)$$

Equation (13) corresponds to the visual second-order rheological model
(Fig.3). The behaviour of the above model is described by the conventional
methods | 7 | using equation

$$P + \left(n_1^k + n_2^k \right) \frac{dP}{dt} + n_1^k n_2^k \frac{d^2P}{dt^2} = 3 \left[K_1 \Theta + \right.$$

$$+ \left[(K_1 + K_3) n_2^k + (K_1 + K_2) n_1^k \right] \frac{d\Theta}{dt} + (K_1 + K_2 + K_3) n_1^k n_2^k \frac{d^2\Theta}{dt^2} \right] \qquad (14)$$

where $n_1^k = \eta_1^k / K_2$, $n_2 = \eta_2^k / K_3$.

A comparison between coefficients of equations (13) and (14) gives us the
relationships sought and represented in Fig. 3.
Models and relationships between rheological parameters for uniaxial loading
(Fig.3) are found by analogy. Equation (5) for transverse strain coefficient
written in operator form taking into account (3), (5) and (9), (11)
will take the form:

$$\hat{\mathcal{V}} = \frac{\hat{M} - 2\hat{G}}{2(\hat{M}-\hat{G})} = \frac{M_\infty \bar{D} \bar{P} - 2 G_\infty \bar{Q} \bar{C}}{2 (M_\infty \bar{D} \bar{P} - G_\infty \bar{Q} \bar{C})} =$$

$$= \frac{M_\infty - 2G_\infty + \left[(M_1-2G_\infty) n_1^M + (M_\infty-2G_1) n_1^G \right] \frac{d}{dt} + (M_1-2G_1) n_1^M n_1^G \frac{d^2}{dt^2}}{2 \{ M_\infty - G_\infty + \left[(M_1-G_\infty) n_1^M + (M_\infty-G_1) n_1^G \right] \frac{d}{dt} + (M_1-G_1) n_1^M n_1^G \frac{d^2}{dt^2} \}} \qquad (15)$$

The transverse strain coefficient cannot be matched with the rheological model
as the coefficient itself is the relationship between bilinear operators (5),
and every operator is described by its own rheological model.
Equation (15) yields values of practical importance for continuous transverse
shear strain $\mathcal{V}_\infty$ equal to the ratio between free members in the numerator and
the denominator (15), instantaneous ν_0 , equal to the ratio between coefficients
with maximum derivatives in the numerator and the denominator (15), coefficient
of the transverse flow strain ν_e equal to the coefficient ratio with the first
derivatives in the numerator and the denominator (15):

$$\mathcal{V}_\infty = \frac{M_\infty - 2G_\infty}{2(M_\infty - G_\infty)} \quad , \qquad \mathcal{V}_0 = \frac{M_1 - 2G_1}{2(M_1 - G_1)} \qquad (16), (17)$$

$$\mathcal{V}_e = \frac{(M_1 - 2 G_\infty) n_1^M + (M_\infty - 2 G_1) n_1^G}{2 \left[(M_1 - G_\infty) n_1^M + (M_\infty - G_1) n_1^G \right]} \qquad (18)$$

Their variations in time are plotted in Fig. 9.
When other pairs of tests are carried out according to Table 1 the procedure of
judgement is analogous to that explained above and is not repeated here.

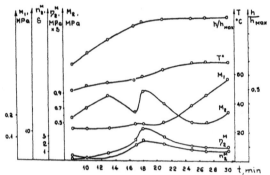

Fig. 4. Changes in rheological parameters of gas concrete mix during swelling under one-dimensional loading. h/h_{max} - swelling curve, T° - temperature curve.

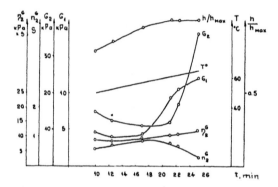

Fig. 5. Changes in rheological parameters of gas concrete mix during swelling under shear.

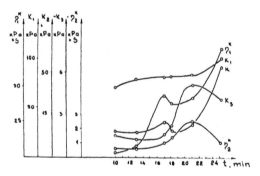

Fig. 6. Changes in rheological parameters of gas concrete mix during swelling obtained by calculation for three-dimensional loading.

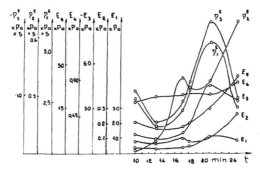

Fig. 7.   Changes in rheological parameters of gas concrete mix during
swelling, obtained by calculation for uniaxial longitudinal
loading.

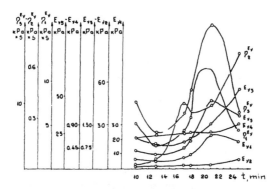

Fig. 8.   Changes in rheological parameters of gas concrete mix during
swelling, obtained by calculation for transversal strain
under uniaxial loading.

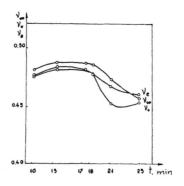

Fig. 9.   Calculated Poisson's ratios (coefficients of transversal
strain) for gas concrete mix during swelling.

101

In this way by selecting rheological models for cases of spherical and uniaxial
loadings (Fig. 3) and by expressing the parameters of the above models as
well as parameters of the transverse strain coefficients by means of rheological
models of shear and one-dimensional loading we have calculated the required
parameters and their time variations (Figs. 6, 7 and 8) from experimental data
given in Fig. 4 and 5.

6 A note

It should be remarked here that the method of operator replacement causes certain
problems with the order of the equation becoming higher. Problems arise in
selecting the rheological model according to the given rheological state equation
It is because mathematical and consequent discussion of types and equations of
rheological models involving more than four elements was not found in literature
The growing number of elastic and viscous elements results in an increasingly
higher number of model types and consequently that of equations. Therefore it is
not easy to find an adequate rheological model even for a second-order equation
and even more so for the above mentioned third-order equations to be matched with
models containing 8 elements.

7 Conclusions

1. Interrelationships for elastic and viscous parameters of linear isotropic
viscoelastic media under four types of uniform loading have been established.
2. The established relationships are of highly practical value as it is enough
to know two of the six sets of values to determine the whole group,namely, two
tests (or even one) are sufficient depending upon the availability of the
equipment or the properties.
3. Two types of tests enable us also to obtain all the coefficients of the
generalized Hooke's law for linear viscoelastic media, i.e., to arrive at the
law itself on scientific grounds and to apply it in solving practical three-
dimensional problems in material processing.
4. It has been found that the selection of the two initial rheological models
strictly determines the degree of the models sought and is non-strict with
respect to their form. In a general case models referring to various types of
loading do not coincide.
5. Shown in the example of a gas concrete mix is the practical implementation of
the method to determine the complete set of elastic and viscous parameters for
linear elastoviscous body. By test and calculation valid data have obtained for
six sets of rheological parameters of the mix and their variations in time.
Simple shear and compression loading were selected as the initial types of load

References

1. Sokolnikoff I.S.,Specht R.O. Mathematical Theory of Elasticity,New York, 194
2. Alfrey T., Jr., Mechanical Behaviour of High Polymers, New York, 1948.
3. Freudental A.M., Geiringer H. The Mathematical Theories of the Inelastic
   Continuum, Berlin, 1958.
4. Месчян С.Р. Механические свойства грунтов и лабораторные методы их определе-
   ния. Москва: Недра, 1974. - 192 с.
5. Ребиндер П.А., Иванова-Чумакова Л.В. Структурно-механические (вязкостно-элас
   тические) свойства растворов полимеров и методы их измерений. В кн.: Успехи
   химии и технологии полимеров. --Сб.2., Москва, 1957. -С. 146-170.
6. Гольберг И.И. Механическое поведение полимерных материалов. -- Москва:
   Химия, 1970. - 190 с.
7. Ржаницын А.Р. Теория ползучести. - Москва: Стройиздат, 1968. - 418 с.

12 RHEOLOGICAL MODEL OF STRUCTURISED SWELLING VISCOELASTIC MEDIUM UNDER COMPRESSIVE DEFORMATION

A. TERENTYEV and G. KUNNOS
Polytechnic Institute of Riga, USSR
L. RUDZINSKI
Technical University of Kielce, Poland

Abstract
Apparatus and methods are described for one-dimensional
deformation (OD) of media in order to be submitted to
rheological analysis. Behaviour of swelling gas concrete
mix under OD conditions is presented. This behaviour is
described by means of a new six-parameter, non-linear
rheological model of structurised swelling viscoelastic
(SSVE) medium. Kinetics of changes of ten rheological
parameters of gas concrete mix is presented in the
processes of swelling and initial structurisation.
A physical interpretation of the behaviour of the selected
parameters is provided.
Keywords: Swelling Viscoelastic Medium, Compressive
Deformation, Gas Concrete Mix, Rheological Model.

1 Introduction

In the shock or vibration technologies of production of
gas concrete blocks, mixture contained in a mould
undergoes a periodical one-dimensional deformation (OD)
caused by the effect of longitudinal waves. Determination
of rational regimes of shock formation based on a
theoretical description of wave processes in mixture is
not possible without knowledge of its rheological
parameters under OD conditions. For that purpose apparatus
has been constructed, and methods of rheological
investigation of mixture in swelling processes and initial
structurisation are described.

2 Apparatus

The design of the apparatus is adapted to the investigation
of materials which quickly change their volume and have low
carrying capacity, e.g. foams. The apparatus also allows
to apply loads which extend within the limits of adhesion
of mixture with a piston.

The investigation cycle of gas concrete mixture
includes: instantaneous loading of material with a constant
load, maintainance of material for some time under load
and instantaneous unloading. Since the material changes
its properties continuously with time lapse, the adopted
cycle was repeated periodically throughout the whole
investigation time.

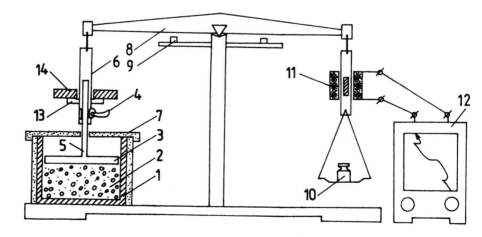

Fig.1. Apparatus for measuring one-dimensional deformation
of swelling medium

The apparatus (Fig.1) operates in the following way.
A stiff metal form 1 is filled with gas concrete mixture 2.
Next piston 3 is introduced to the mould until it has a
full contact with mixture surface. The cam-lever grip 4
is in open position, thanks to which the piston shank 5
can loosely travel within holder 6. In order to reduce
heat excretion from mixture to the atmosphere, the form is
covered with a thermoinsulating lid 7. Position of the
piston rod in the grip is fixed with the help of clamp 4,
the left arm of the scales beam 8 being earlier pressed to
limiter 9. In this way vertical displacements of the
piston and its effect on mixture are eliminated. When, in
turn the swelling mixture presses the piston, it moves
freely in the mould 1 together with the surface film of
adherent mixture because it is counterbalanced with load
10. These displacements are recorded by means of an
induction sensor 11, and the recorder 12 plots the curve
of mixture swelling under zero load.
Instantaneous load of material is accomplished by means
of a wide disk 13 on which vessel bobs 14 are laid. After
10 - 30 seconds of material swelling under load, bobs are
quickly removed and material is unloaded.

3 Phenomenological behaviour of gas concrete mixture

The swelling medium is in the state of constant deformation which is described in the coordinate system $\varepsilon = f(t)$ by a certain curve of swelling "a" (curve l/l_{max} in Fig.6), its angle of inclination to the time axis $< \alpha$ changes continuously. Under pressure P the medium swells according to a certain curve "b", whose angle of inclination to the time axis $< \beta$ decreases continuously in an analogous way, and at a certain moment during swelling $\alpha > \beta$ when $P < 0$ and $\alpha < \beta$ when $P > 0$; besides, the greater $|P|$, the greater $|\alpha - \beta|$.

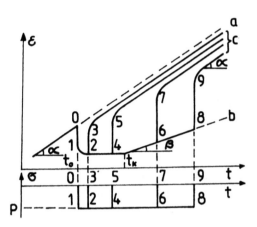

Fig.2. SSVE medium (gas concrete mix) deformation vs effect time of external pressure

Since the swelling time is much longer than that of load effect, considering each cycle of load separately, the curves of the swelling line can be approximated with straight lines "a" and "b" whose angles of inclination α and β in a given time interval are constant (Fig.2). It is assumed at the same time that the rheological properties of material are invariable within the limit of one cycle of load.

In this way material deformation is added increasingly to the load moment with a constant rate defined as tgα. An instantaneous deformation of material is observed at momentary load according to the straight line 0 - 1. Next there occurs retarded deformation which after reaching a certain level becomes stabilized and forms the so-called "shelf" - the first characteristic effect. At moment tk the deformation rate changes from 0 to tgβ, and material is deformed once again creating a straight line of swelling "b" under load. Instantaneous unloading occurs at point 8 (Fig.2) during which material is momentarily deformed according to the straight line 8 - 9, and then follows retarded deformation which is finally transformed into the

straight line of material swelling without loading "c"
being parallel to the straight line "a".

It should be noted that the second characteristic effect
- instantaneous deformation under mixture unloading exceeds
deformation under loading. If unloading is performed
behind point tk, i.e. when deformation proceeds to the
straight line "b" practically identical unloading curves
are obtained. If unloading is performed on the "shelf" up
to the value tk, a decrease in instantaneous deformation
is observed along with a decrease in the time interval
between loading and unloading of material. Respectively
swelling straight lines "c" after unloading remain parallel
to the straight line "a" after loading (Fig.2).

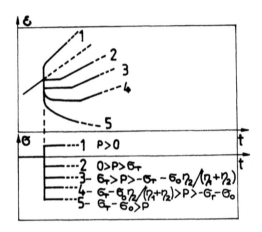

Fig.3. SSVE medium deformation vs time

In the experiment also the magnitude of static loading
was changed, its influence on the shape of the deformation
curve is represented in Figure 3. Under small compressive
loads of type 2 along the segment of instantaneous loading
there is no instantaneous deformation of material. Under
slightly greater loads of type 3 instantaneous deformation
is observed; however, it is connected under a right angle
with a segment of deformation stabilization. Under much
greater loads of type 4 instantaneous deformation is
followed by retarded deformation which ends with a stabilization segment. In the considered three cases of loads
the length of the stabilization segment of deformation
grew with an increasing load, whereas the inclination angle
of the straight lines of material swelling decreased. Under
considerable compressive loads of type 5 material lost its
load capacity and was deformed irreversibly.

After applying tensile loading 1 to material (Fig.3) an
increase in the mixture swelling rate was observed after
its instantaneous deformation.

4 Rheological model of swelling gas concrete mixture

The described effects result from swelling of gas concrete mixture under one-dimensional loading. To describe behaviour of swelling gas concrete mixture a new rheological element is introduced, which symbolizes a source of internal pressure in material and leads to swelling of system. This element represents a cylinder with weightless piston, moving into cylinder without friction. The constant pressure is kept up within the cylinder independently of piston displacement. The element, joined parallelly with viscous element, can not appear individually because the piston would push out at once by internal pressure. This way swelling is described by means of the following equation:

$$\sigma_0 + \sigma(t) = \eta \, \dot{\varepsilon}_1(t) \tag{1}$$

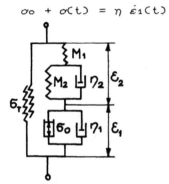

M_1, M_2 – OD instantaneous and retarded elastic moduli,
η_1, η_2 – OD viscosities,
ε_1, ε_2 – OD strains,
σ_0 – swelling pressure
σ_T – limit stress of the SSVE medium frame stability (at tensile deformation $\sigma_T = 0$),
$H = M_1 \cdot M_2 / (M_1 + M_2)$ – OD long term modulus of elasticity,
$n_1 = n_2 H / M_1$, $n_3 = \eta_1 / M_1$ – relaxation times,
$n_2 = \eta_2 / M_2$ – retardation time

Fig. 4. SSVE medium rheological model

Finally, the rheological model has been proposed in Fig. 4. Because ratchet mechanism is an element of one-dimensional action so model equations will differ during compression and unloading. In the process of compression when ratchet mechanism works and when external pressure is equal to $P(t)$, the stress in right chain is described as:

$$\sigma(t) = P(t) - \sigma_T \tag{2}$$

Denoting by $\varepsilon_2(t)$ a part of the total strain, corresponding to deformation of Kelvin-Voigt model, and by $\varepsilon_1(t)$ -deformation of swelling element of model, the total deformation is described as:

$$\varepsilon = \varepsilon_1(t) + \varepsilon_2(t) \tag{3}$$

and Kelvin-Voigt equation is:

$$\sigma(t) + n_1 \dot{\sigma}(t) = H\varepsilon_2(t) + M_1 n_1 \dot{\varepsilon}_2(t) \tag{4}$$

After eliminating $\varepsilon_1(t)$, $\varepsilon_2(t)$ and $\alpha(t)$ from system of equations (1), (2), (3), and (4) an equation of model of SSVE medium under compressive load $P(t)$ is given by:

$$\ddot{P}(t) + \left(\frac{1}{n_1} + \frac{1}{n_3} \right) \dot{P}(t) + \frac{P(t) + \sigma_o + \sigma_T}{n_2 n_3} = \frac{H}{n_1} \dot{\varepsilon}(t) + M \ddot{\varepsilon}(t) \qquad (5)$$

All presented regularities are described by means of the equation (5) (Figs. 2 and 3). Besides the intervals of changes in external pressure according to rheological parameters of medium are determined for cases, shown in Figure 3.

Analytic solution of equation (5) for investigated way of loading gives a possibility to determine rheological parameters of gas concrete mixture basing on the experimental deformation curves $u(t)$ (Fig. 5).

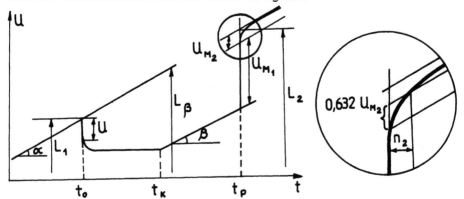

Fig. 5. Schematic diagram to determine rheological parameters in model of SSVE medium

$$M_1 = \frac{P}{\Delta \varepsilon_{M_1}(t_p)} = \frac{P \, L_2}{u_{M_1}} \,, \qquad M_2 = \frac{P}{\Delta \varepsilon_{M_2}(t)} = \frac{P \, L_2}{u_{M_2}} \,;$$

$$\sigma_T = [\Delta \varepsilon_{M_1}(t_p) - \Delta \varepsilon(t_o)] \, M_1 = \left(\frac{u_{M_1}}{L_2} - \frac{u}{L_1} \right) \circ M_1 \qquad (6)$$

$$\eta_1 = \frac{P}{\dot{\varepsilon}_\alpha - \dot{\varepsilon}_\beta} = \frac{P \, L_1 \, L_\beta}{L_\beta \, tg_\alpha - L_1 \, tg_\beta} \,; \qquad \sigma_o = \frac{tg\,\alpha}{L_1} \eta_1$$

For $t = t_p + n_2$ an equation of delayed deformation under unloading is given by:

$$\varepsilon(t) = \frac{\sigma_o}{M_1} n_2 + \varepsilon(t_p) + \frac{P}{M_1} + \frac{P}{M_2} \, 0,632 \qquad (7)$$

Retardation time n_2 can be determined using graphical method (insert in Fig. 5).

5 Experimental and discussion

All 6 parameters of gas concrete mixture together with 4
derivative parameters and their transformations within 30
minutes after addition of water to solid mixture were
determined on the basis of the above presented method of
measurement (Fig. 6).

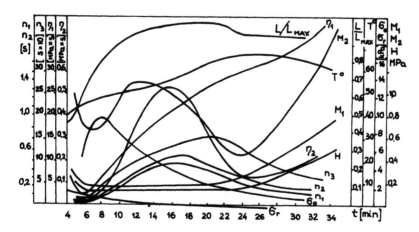

Fig. 6. Changes of rheological parameters of gas concrete
mixture in the processes of swelling and struc-
turization; L/Lmax is a swelling curve, T^o is a
temperature change

At first we shall consider the source of internal
pressure (element σ_0 in Fig. 4), which symbolizes the work
of gas evolution inside the mixture during the reaction of
aluminium powder with alkalis. Every aluminium particle is
covered with an oxygen film which hampers the coarse of a
given chemical reaction. Therefore, this reaction occurs
only in places where the film is cracked or damaged under
the influence of alkalis. Chemical destruction of oxygen
films develops gradually, temperature having a significant
influence on the rate of chemical reactions. Curve increase observed up to 8th minute in the courses of curve
σ_0 in the function of time (Fig. 6) can be explained by an
increase in the surface of the reaction area of gas evolution and temperature growth as a result of occurring
chemical reactions of lime hydration. Until 8th minute
mixture swelling reached 65% of its maximal magnitude, i.e.
more than half of the gas-forming agent reacted. Only
coarser aluminium particles and particles with durable
oxygen film remained in mixture. Therefore, after that
time, irrespective of temperature increase, the intensity
of gas evolution decreases as a result of the continuously
decreasing mass of the gas-forming agent in mixture.

109

Changes in viscosity η_1 due to the evolution of gas inside the mixture force it to deform, and finally, to swell. Swelling rate depends mainly on the viscosity of the interporous substance and its resistance to expansion and renewed formation of gas bubbles. Therefore, parameter η_1 can be referred to the viscosity of the interporous substance where the physical sence of viscous resistance consists in the displacements of constant particles, their mutual friction, leakage, and percolation of liquid components of mixture with viscous resistance. From the moment of application of water the interporous substance contains bridges of chemical compounds of the first weak structures, after which together with increasing temperature the process of structurisation becomes intensified and, apart from it, as a result of hydration reaction the amount of free water, which plays the part of lubricant between solid particles, continuously decreases in interporous substance. All this contributes to the increase in viscosity with time, which can be observed in the experiment (Fig. 6).

Furthermore, we shall consider from the physical point of view deformation of gas concrete mixture under load (changes in parameter σ_T). Mixture is a theree-phase system in which each component has specific deformation properties. Under small loads mixture compressibility is defined as a rule by the compressibility of the gas component - gas bubbles, steam and air. Under the static loading of swelling mixture, this loading is initially taken over by the structural skeleton composed of the solid components of mixture. Since the employed loads are sufficiently small to fully neglect the volume compressibility of liquid and solid components of mixture, loading should lead to the volume compression of gas bubbles. The bubble area should decrease which requires displacement of solid particles and liquid. This displacement should be parallel to the friction of particles between one another and disruption of certain microstructural bonds, which means that in order to overcome such forces it is necessary to have a certain limit stress P_T which characterizes the strength of the structural bonds of material.

In our investigation magnitude σ_T, calculated from formula (6), essentially should have defined the difference P_T between external pressure and stress in gas bubbles at the moment of loading. However, formula (6) includes the difference of instantaneous deformations under loading and unloading of material. With time lapse the amount of free steam in interporous substance decreases and this substance is transformed from concentrated solid particles suspended in liquid into a visco-elastic-plastic body which is characterized by plastic strains under unloading. As a result instantaneous deformation under unloading decreases, which leads to a decrease in σ_T with time lapse until negative values are reached. This is evidence of a full domination

of the effect of plastic deformation development (Fig.6).
In this way the behaviour of σ_T in time is defined with the
help of two different effects. Sum σ_0 and P_T means mixture
bearing strength. When it is exceeded irreversible settlement occurs.

6 Conclusion

1. The results of the investigation have revealed specific
properties of the compressive deformation of swelling gas
concrete mixture as a multifunctional component; particu-
lary instantaneous deformation under unloading exceeds
instantaneous deformation under loading and there is a
certain segment of deformation stabilization after instantaneous loading of SSVE medium by constant pressure. This
effect is explained by a gradual egualization of stresses
between the skeleton of SSVE medium and pore liquid with
gas bubbles as a result of gas evolution after loading the
medium by external pressure.
2. The rheological behaviour of swelling gas concrete mix
under one-dimensional deformation can be described by a
6-parameter model taking into account the elastic, viscous,
and structural properties of medium and also containing a
new, active rheological element - the source of internal
pressure.
3. The work of the model is described by a differential
equation of the second degree, the solution of which under
the investigated material loading and unloading allows to
describe all experimentally determined regularities
concerning deformation of swelling gas concrete mixture.
4. The apparatus and method applied allow to obtain
experimental data taking into account the simultaneous
change in time of all 6 parameters of the model of swelling
gas concrete mixture of a given composition.

7 References

Citovich, N.A. (1973) Mechanika Gruntov. Vysshaja Shkola,
    Moscow.
Dresher, A. and Jons, G. (1972) Photoelastic verification
    of a mechanical model for the flow of a granular
    material. J.Mechanics and Physics of Solids, 20, 337
    -357.
Kunnos, G.J. (1978) Reologiczeskije modeli tela s razlicz-
    nym poviedenijem pri nagruzenii i razgruzkie, in Tech-
    nologiczeskaja Mechanika Betona, Polytechnic Institute
    of Riga, pp. 55-73.
Lyakhow, G.M. (1982) Volny v Gruntach i Poristych Mnogo-
    komponentnych Sredach. Nauka, Moscow.
Mironov, V.E. Kunnos, G.J. and Khoromieckij, V.G. (1979)
    Uprugovyazkoplasticzeskije svojstva gazobetonnych

smesiej pri vibrovspuczivanii, in **Technologiczeskaja Mechanika Betona,** Polytechnic Institute of Riga, pp. 33-54.

Nikolayevskij, V. N. Basniyev, K. S. Gorbunov, A. T. and Zotov, G. A. (1970) **Mechanika Nasyszczonnych Poristych Sred.** Nedra, Moscow.

Sukle, L. (1976) **Rheological Aspects of Soil Mechanics** (in Russian). Stroyizdat, Moscow.

Terentyev, A. E. (1984) **Volnovyje Processy vo Vspucziva-juszczejsia Uprugovyazkoj Sredie pri Vertikalnych Udarnych Vozdejstvijach.** Ph D Thesis, Polytechnic Institute of Riga.

Terentyev, A. E. and Kunnos, G. J. (1988) The reology of the structurised swelling viscoelastic medium with thixotropy, in **Proceedings of the Xth International Congress on Rheology,** University of Sydney, pp. 317-319.

13 MODELLING IN ANALYSIS OF DEFORMATION OF VERY EARLY AGE CONCRETE SUBJECT TO CONSTRUCTION LOAD

H. OKAMOTO
Maizuru College of Technology, Maizuru, Japan
T. ENDOH
Central Research Institute of Electric Power Industry,
Abiko, Japan

Abstract
The knowledge of deformation of very early age concrete subject to
construction load has become very important to introduce
systematization and robotization in concreting practice. Study on
structural analysis of very early age concrete, has not been fully
established yet because mechanical properties show very complicated
behavior during the setting process of cement. For the purpose of the
rationalization of concreting practice, the present study has been
carried out to obtain knowledge on deformation of concrete for
construction load resulted from the stripping, and to evaluate physical
properties of setting process with rheological model. In present study,
we propose that deformation of very early age concrete subject to
sustained load can be estimated both with finite element analysis and
with simple analysis.
Keywords: Creep, Rheological model, Visco-elasticity, Slip-form,
Setting process, Deformation, Finite element analysis, Simple analysis.

1 Introduction

For the setting process on concrete, it is recognized that although
studies to make clear character on cement have been extensively carried
out in cement chemistry, studies on physical properties are few in
concrete engineering, because too much emphasis is put to obtain an
index of construction control in concreting practice. The setting
process of concrete indicates very complicated properties because of
the change from fluid state into solid state caused by the progress of
hydration of cement and that of sedimentation of granular materials.
    Recently, concreting practice is necessary to introduce
systematization and robotization on account of of the employing
situation getting worse and works growing large-sized. The knowledge
of deformation of very early age subject to construction load
which consists of dead load and working machine parts after the
stripping of form has become very important in slip-form construction
and precast concrete products of instant stripping.
    The present study has been carried out to obtain knowledge on
deformation resulted from the stripping of form in work, and to
evaluate rheologically the properties of the setting process with the
constants by visco-elasticity. The following items are investigated
for the present study; (1) To establish constitutive equation of very

early age concrete applying rheological model. (2) To evaluate the pysical properties of the setting process with rheological model. (3) To apply finite element method to analyze deformation. (4) To present simple nalysis applying theory of linear elasticity. (5) To compare measured results with calculated results on the problem of deformation by dead load of column.

2 Constitutive equation

2 . 1 Rheological model

The problem on deformation of very early age concrete subject to construction load has been required to consider creep behavior by sustained dead load resulted from stripping of form in slip-form construction. As for rheometer for the purpose of evaluation rheologically on the setting process, creep test is most suitable because time-history of strain can be measured and with it visco -elasticity can be expressed. But, the study of creep on very early age concrete is unsatisfactory up to now, compared with that on hardened concrete.

For constitutive equation of creep on very early age concrete, it is suitable to apply rheological model by the combination of visco -elasticity as follows. In applying three-element model as shown Fig. 1 (a), total strain ε is given by the following function,

$$\varepsilon = \varepsilon_e + \varepsilon_f = \frac{\sigma}{E} + \frac{\sigma}{E_I} [1 - \exp \{-\frac{E_I}{\eta_I} t\}] \qquad (1)$$

where, ε_e; the instantaneous strain, ε_f; the creep strain, σ; the stress by sustained loading, E; the instantaneous modulus of elasticity, E_I; the delayed modulus of elasticity, η_I; the delayed coefficient of viscosity, t ; the elapsed time from start of loading.

In case of four-element model as shown Fig. 1(b),

$$\varepsilon = \frac{\sigma}{E} + \frac{\sigma}{E_I} [1 - \exp \{-\frac{E_I}{\eta_I} t\}] + \frac{\sigma}{\eta} t \qquad (2)$$

where, η; the relaxation coefficient of viscosity.

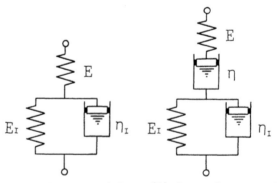

(a) Three-element model   (b) Four-element model
Fig. 1   Rheological models

2.2 Experimental method

Creep test in sustained loading of compression was carried out to estimate physical properties of very early age concrete by using the apparatus as shown in Fig. 2, with cylindrical specimen of $\phi 10 \times 20$ cm. The axial displacement of specimen corresponding to elapsed time of sustained load was measured in creep test. Stripping form of specimen after removing bleeding water immediately before each test, loading test was sustained for 30 minites. The age of test was from 1 hrs to 7 hrs for curing temperature of 10℃ and 20℃. The ratio of sustained loading stress was 0.35 and 0.60 to static strength which was obtained beforehand.

Mix proportion and the measured slump value of concrete are shown in Table 1. Ordinary portland cement with specific gravity of 3.16 and initial setting of 2 hrs 25 min. and final setting of 3 hrs 53 min., river sand with specific gravity of 2.56, and crushed stone with specific gravity of 2.62 were used. As admixture AE agent was used.

Table 1 Mix proportion and slump values

W/C(%)	s/a(%)	Unit weight (kg/m³)			
		W	C	S	G	AE(g)
50	48	190	380	832	926	150
Measured slump values		20℃: 8~10cm			
		10℃: 10~12cm			

2.3 Experimental results and discussion on rheological constants

Total strain ε in creep test is divided into the instantaneous strain and the creep strain. The rheological constants applying the data of creep strain to Eq.(1) in case of 3-element model and Eq.(2) in case of 4-element model were calculated by means of the method of non-linear least squares. Relationship between the logarithm of creep strain and the loading time at the age of 4 hrs is indicated in Fig. 3, comparing the measured values of creep strain with the estimated values. It shows therefore that 4-element model is in approximately better agreement with measured results than 3-element model. Other results also show the same tendency above. Consequently, it is considered that constitutive equation on creep behavior of very early age concrete fits 4-element model well.

Relationship between the age and the rheological constants by 4-element model, is shown in Fig. 4 plotting semilog graph paper. Each rheological constant grows exponentially with the increase of age, and with the rise of curing temparature. Above-mentioned tendency shows properly a change of state depending on progress of hydration of cement and sedimentation of granular merterials including bleeding water.

In relation to the chemical characteristics of ordinary portland cement a behavior on the development of hydration as shown in Fig. 5 is recognized generally. While particles of cement keeps in contact with water, hydration progresses for about ten minutes and then it stops temporarily, and three hours later it progresses again. Consequently, hydration can be divided into three stages as follows; the first stage is initial rapid reaction period, the secondary stage is induction period and the tertiary stage is rate-increasing period. Although hydration of cement stops in the secondary stage, strength and visco -elasticity as concrete grow gradually because cohesion and internal

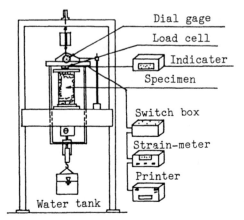

Fig. 2 Apparatus of creep test

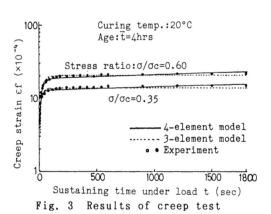

Fig. 3 Results of creep test

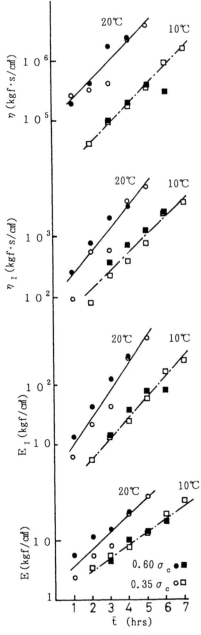

Fig. 4 Experiment results of rheological constants

Fig. 5 Development of hydration of ordinary portland cement

friction by sedimentation of particles make progress.

2 . 4   Creep behavior by multi-stage loading
Considering construction load resulted from overlaying in turn by placement of concrete like in slip-form construction, creep test by multi-stage loading was carried out, in which first loading was at the age of 2.0 hrs, secondary loading was at 2.5 hrs, tertiary loading was at 3.0 hrs, and last loading was at 3.5 hrs, with stress ratio of 0.60, in curing temperature of 20℃. The experimental results are shown in Fig. 6, which gives the relationship between total strain and elapsed time. The strain by the first loading increased largely, but the strain by additional loading after that took extremely small quantity. It is considered the reason is that concrete was consolidated so strongly in the first loading and the surface of specimen dried in creep test, which led to high density.

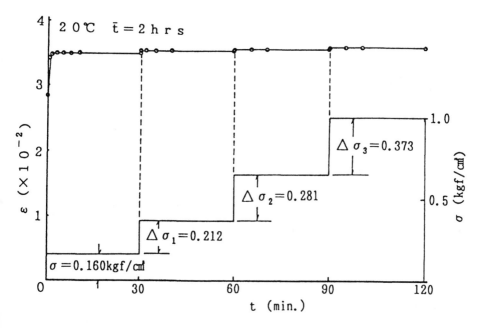

Fig. 6   Creep test result by multi-stage loading

2 . 5   Creep rupture
Creep test under high stress ratio was carried out to investigate the behavior of creep rupture. The representative results of creep rupture are indicated in Fig. 7, under stress ratio of 0.90. Various types of creep rupture were observed. It is considered that creep limit of very early age concrete is about 80%.

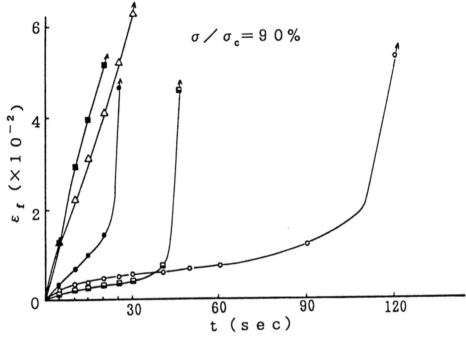

$\sigma / \sigma_c = 9\,0\,\%$

Fig. 7   Representative results of creep rupture

3   Finite element analysis

Calculation of deformation of very early age concrete subject to
construction load applied finite element analysis by 4-element model.
Now, assuming isotropic materials with 4-element model as shown in Fig.
8, the following equations are obtained under three dimensional stress,

$$\{\dot{\sigma}\} = [D_g^e]\{\dot{\varepsilon}\} - \{\dot{A}\} \tag{3}$$

$$\{\dot{A}\} = [D_g^e]([\eta_g]^{-1} + [\eta_i]^{-1})\{\sigma\} - [D_g^e][\eta_i]^{-1}[D_i^e]\{\varepsilon_i\} \tag{4}$$

where, $\{\dot{\sigma}\}$, $\{\dot{\varepsilon}\}$, $\{\sigma\}$; stress velocity, strain velocity and stress of the
entire element, respectively, $\{\varepsilon_i\}$; strain on the part of Voigt-model,
$[D_g^e]$, $[D_i^e]$; stiffness matrix of each spring element, $[\eta_g]$, $[\eta_i]$;
viscosity matrix of each dashpot.
    This analysis need to solve the differential equation on $\{\varepsilon_i\}$, unknown
strain of Voigt element, given by the following formula,

$$\{\dot{\varepsilon}_i\} = [\eta_i]\{\sigma\} - [\eta_i][D_i^e]\{\varepsilon_i\} \tag{5}$$

    The following equations can be used to extend 4-element model under
uni-axial stress to that under three axial stress,

$$E_g = E \tag{6}$$

$$E_i = E_I \tag{7}$$

$$\eta_{Gg} = \frac{\eta}{2(1+\nu_g)} \tag{8}$$

$$\eta_{Kg} = \frac{\eta}{3(1-2\nu_g)} \tag{9}$$

$$\eta_{Gi} = \frac{\eta_I}{2(1+\nu_i)} \tag{10}$$

$$\eta_{Ki} = \frac{\eta_I}{3(1-2\nu_i)} \tag{11}$$

As for further details, it is expected to refer to the previous report of the authers(1988).

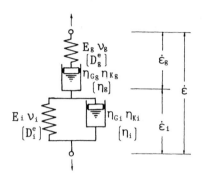

Fig. 8 Generalized 4-element model

4 Deformation by dead load of column

4.1 Experiment

Deformation test on dead load of column was performed to investigate the accuracy of analysis on deformation of very early age concrete subject to sustained load. The outline of the experiment is shown in Fig. 9. The dimension of specimen of column is the section of 20×40 cm with the height of 100 cm. In this experiment, when concrete placed in a form with mix proportion shown in Table 1 was stripped at the age of 3 hrs, vertical displacement of the top and lateral displacement at the height of 40 cm from bottom were measured to determine changes during 30 minutes after stripping by using two cathetometers of digital type with accuracy of 1/100 mm. Creep test was simultaneously conducted to decide rheological constants used for analysis.

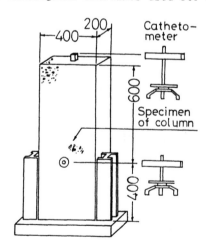

Fig. 9 Deformation test of column

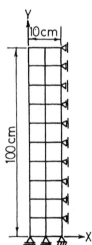

Fig.10 Structural model

119

4. 2 Analysis by FEM

Numerical simulation on the above-mentioned problem was carried out by finite element method mentioned in the foregoing paragraph with the mesh-divided structural model as shown in Fig.10 using tetrahedral element of 8 nodal points. The time-history of deformation was calculated assuming that dead load of concrete acts at the time of $t=0$, considering the state of plane strain. Boundary condition in the bottom of specimen was investigated by roller end. Rheological constants used for analysis are indicated in Table 2 on the basis of Fig. 12. Poisson's ratio is assumed $\nu_g = \nu_i = 0.45$, as investigated by Byfors (1980) for strength below 1.0 kgf/cm².

Table 2 Rheological constants
used for analysis(age of 3 hrs)

E	7.88kgf/cm²
E_I	51.5kgf/cm²
η_I	3.51×10^3 kgf·sec/cm²
η	2.34×10^5 kgf·sec/cm²
$\nu = \nu_g = \nu_i$	0.45
ρ	2.33g/cm²

4. 3 Simple analysis by linear elasticity

Simple analysis by linear elasticity is suggested on deformation by dead load of column as follows. Column as shown in Fig.11 is dealt with as the state of plane strain, assuming isotropic materials. Relationship between stress σ and strain ε in this state is represented by the following equation,

$$\varepsilon_x = \frac{1}{E_e}\{(1-\nu^2)\,\sigma_x - \nu(1+\nu)\,\sigma_y\} \tag{12}$$

$$\varepsilon_y = \frac{1}{E_e}\{(1-\nu^2)\,\sigma_y - \nu(1+\nu)\,\sigma_x\} \tag{13}$$

$$\gamma_{xy} = \frac{1}{G}\tau_{xy} \tag{14}$$

In this problem, $\sigma_x = 0$, the stress in the position y,

$$\sigma_y = \rho(H-y) \tag{15}$$

Average stress in the direction of y is given by

$$\bar{\sigma}_y = \frac{1}{H-y}\int_y^H \sigma_y dy = \frac{1}{2}\rho(H-y) \tag{16}$$

Thus vertical displacement δ_y and lateral displacement δ_x are given by

$$\delta_y = \frac{1}{E_e}\{(1-\nu^2)\,\bar{\sigma}_y\}y \tag{17}$$

$$\delta_x = \frac{1}{E_e}\{\nu(1+\nu)\,\sigma_y\}b \tag{18}$$

where, ρ; density, and E_e; equivalent elastic modulus. E_e is shown as

$$E_e = [\frac{1}{E} + \frac{1}{E_I}\{1-\exp(-\frac{E_I}{\eta_I}t)\} + \frac{1}{\eta}t]^{-1} \tag{19}$$

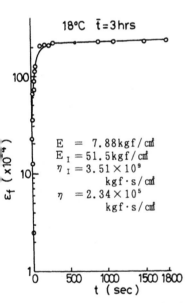

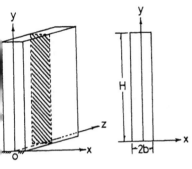

Fig.11 Plane strain in column

$$E = 7.88 \text{kgf/cm}^2$$
$$E_I = 51.5 \text{kgf/cm}^2$$
$$\eta_I = 3.51 \times 10^9 \text{kgf·s/cm}^2$$
$$\eta = 2.34 \times 10^5 \text{kgf·s/cm}^2$$

18°C $\bar{t} = 3$ hrs

Fig.12 Result of creep test used for analysis

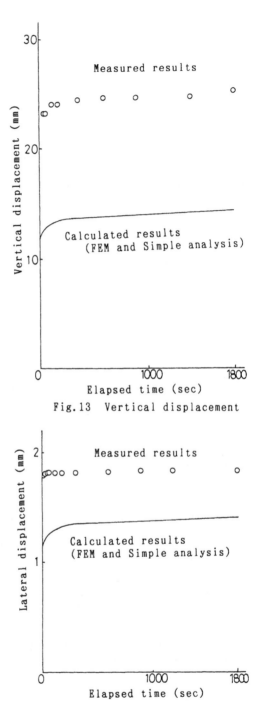

Fig.13 Vertical displacement

Fig.14 Lateral displacement

4. 4 Comparison between experiment and analysis
The result of creep test is indicated in Fig.12, which decided
rheological constants for analysis. The time-history of vertical
displacement and of lateral displacement, comparing the measured
results of deformation with the calculated results, is shown in Fig.13
and Fig.14, respectively. Both the vertical displacement and the
lateral displacement increased by a certain amount just after the
stripping, and later as time elapsed they increased gradually, and
still later they are settled down to a certain value.
   Calculated results by finite element analysis and simple analysis
corresponded with vertical and lateral displacement. Calculated
results increased more highly than measured results, but both of the
time-history curves showed almost the same shape. It is considered
that the difference between measured values and calculated values
resulted from the difference of deformation just after the stripping,
because the reliability of the instantaneous modulus of elasticity E,
which was used for this analysis, was inferior to observational error
of instantaneous strain in creep test.

5   Conclusions

Conclusions obtained from the present study are as follows:
(1) On very early age concrete in sustained loading, constitutive
equation based on four-element model agrees with experimental results
in creep test.
(2) As the age and the temperature of concrete increased, rheological
constants by four-element model at very early age increased
exponentially. Properties of the setting process of very early age
concrete can be evaluated by visco-elasticity by means of creep test.
(3) On creep by multi-stage loading, the strain due to first loading
increased largely, but the strain due to additional loading was
extremely small in quantity.
(4) As stress ratio became more than 0.80, creep rupture occurred.
(5) On deformation by dead load of column, although calculated results
by finite element analysis and simple analysis corresponded, measured
results increased higher than calculated results, because the
reliability of the instantaneous modulus of elasticity by creep test
was unsatisfactory owing to by observational error, but both of the time
-history curves showed almost the same shape. Inputting reasonable
rheological constants, deformation of very early age concrete subject
to construction load can be estimated by analysis presented with this
study.

Acknowledgement
The authors wish to thank Dr J. Murata, Emeritus Professor of Tokyo
Metropolitan University, for his helpful advices.

References
Okamoto, H. and Endoh, T. (1988) A study on foreknowledge of
   deformation of very early age concrete subject to sustained load,
   Proceedings of Japan Society of Civil Engineering, No.396 V -9, pp.69
   -77(in Japanese).
42-CEA Committee (1980) Properties of concrete at early ages,
   Materials and Structures, Vol.13, No.75, pp.265-274.

PART THREE

PROPERTIES OF OILWELL CEMENT SLURRIES AND CEMENTING PROCESSES

14 USE OF A CONTROLLED STRESS RHEOMETER TO STUDY THE YIELD STRESS OF OILWELL CEMENT SLURRIES

P.F.G. BANFILL and D.R. KITCHING
University of Liverpool, Liverpool, UK

Abstract
This paper describes measurements of the yield stress of oilwell cement slurries obtained using a controlled stress rheometer with parallel plate, vane in cup and annular plate and cone geometries. The latter was found to be the most satisfactory and differs from a conventional cone and plate by the provision of a recess in the plate under the cone to give a minimum gap of 1mm while ensuring constant shear rate across the gap. The computerised stress control on the rheomter was programmed to allow a known shear energy to be input to the slurry before each determination of yield stress. At ambient temperature and pressure, yield stress decreases with increasing water content and mixing energy but increases with standing. Slippage is reduced when the surfaces of the measuring systems are roughened.
Keywords: Yield stress, Oilwell cements, Controlled stress measurements, Mixing energy, Shear history.

1 Introduction

Oilwell cement slurries are pumped down the well bore and up the annular space between the casing and surrounding strata where they harden to form a strong, supporting and impermeable seal. The rheology of the slurry is important to ensure easy pumping and complete filling of the annulus without excessive separation of water and cement. The slurry is generally considered to conform to the Bingham model possessing a yield stress (often referred to in the oil industry as gel strength) which is influenced by concentration, previous shear history, time and temperature. In addition the increase of yield stress with time when the slurry is static for any reason influences the ease with which pumping can be restarted. For these reasons it is important to be able to measure the yield stress.

Industrial methods of preparing and testing slurries whether for planning a cementing job or for quality control are well established (API, 1986). However, while simple, they may be criticised on two grounds. Firstly, the prescribed slurry preparation procedure using a Waring blender puts in much more mixing energy to the sample than would be experienced by a slurry in the field (Orban, et al 1986), i.e. the shear history of the two are different. Secondly, the measurement procedure is based on a Fann viscometer and requires extrapolation to zero shear rate of the best straight line through several fixed speeds.

In principle controlled stress rotational rheometry offers more reliable way of determining the yield stress than conventional controlled speed rotational viscometry, because shear stress can be measured at the point of first shearing th slurry instead of extrapolating back to zero shear rate. This paper describes work done on using the CarriMed CS Rheometer f this purpose, with the objective of studying the effect of shearing energy input to a sample of slurry prior to testing i

2 Apparatus

2.1 Rheometer

The CarriMed CS rheometer consists of a stationary, temperatur controlled bottom plate, which is adjustable vertically by a micrometer, and a central shaft mounted on an air bearing whic is driven by an electronically controlled induction motor. A variety of measuring systems - cones, discs, cylinders and van - may be mounted on the shaft and, when operating, the drive motor applies a controlled torque to it which stresses the sample and causes it to flow. The resulting rotation of the shaft is monitored optically. The entire instrument may be programmed externally and the data fed to a computer - in our case an Apple IIe. The flexibility of the instrument is limit only by the flexibility of the software and in the standard mo for determining a flow curve, the torque is increased in 200 steps up to the chosen maximum and at a ramp rate selected and controlled by the software. The resulting angular motion may recorded in radians for small deflections or in radians/sec fo calculating shear rates. The standard software may be modifie for special purposes. The yield stress may be taken from the torque at which flow first occurs and the whole flow curve may be fitted to a chosen model by data analysis software.

2.2 Measuring geometries

Three measuring geometries were used in this work:
  parallel plate, considered in detail by Krieger and Woods (1966);
  vane in cup, described by Nguyen (1983);
  annular plate and cone (APC), after the annular cone described by Cross and Kaye (1986, 1987).
These are sketched in Figs. 1 and 2 which also show the stationary lower plate in the rheometer.

The radius, R_p, of the parallel plate was 25mm and the gap height, h, of 1.0mm was chosen to be 10 times the largest particle diameter expected in a cement slurry. The vane had

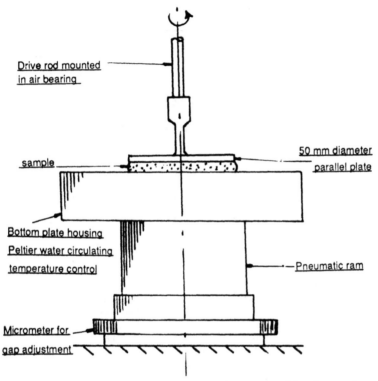

Fig.1. Layout of parallel plate on CS rheometer.

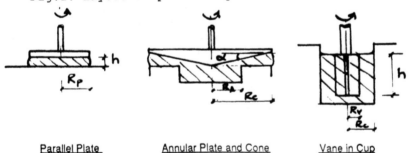

Parallel Plate   Annular Plate and Cone   Vane in Cup

Fig.2 Notation for conversion formulae

four rectangular blades of radius, Rv, 6mm, and height, h, 24mm
in a cup of radius, 15mm centrally mounted on the lower plate.
Because of the restricted size of the rheometer these dimensions
do not comply with those recommended by Nguyen. In the APC
geometry a standard cone of radius, R_c, 30mm and angle,α, 4° was
used in conjunction with a plate with a central section of
radius, R_a, 15mm which was recessed 10mm deep. The vertical
position of the annular plate was set so that the vertex of the

Table 1.   Conversion formulae.

System	Shear stress	Shear rate
Parallel plate	$\tau = 3T/2\pi R_p^2$	$\dot{\gamma} = 3R_p\,\Omega/4h$
APC	$\tau = 3T/2\pi(R_c^3 - R_A^3)$	$\dot{\gamma} = \Omega/\tan\alpha$
Vane in cup	$\tau = 3T/2\pi(R_v^3 + 3R_v^2 h)$	—

cone was in the same plane as the unrecessed section of the
annular plate, and this gave a minimum gap width of 1.05mm.
formulae for calculating shear stress and shear rate from tor
T and angular velocity Ω are given in Table 1. All three
geometries are capable of providing data on yield stress as
defined in Section 2.1 and it would be expected that the vane
would be the most reliable because the cement slurry sample i
sheared within itself rather than at the surface of the movir
member. However, to achieve the objective of providing a
quantifiable energy input during the course of testing, the
shear rate must be known accurately. This is not possible wi
the vane because the position of the sheared plane and hence
dimensions of the sheared zone are not known accurately. In
parallel plate the shear rate varies radially across the samp
Therefore only the APC offers the combination of reliable yie
stress determination and a known shear rate which is the same
throughout the sample test annulus. Finally, as Mannheimer
confirmed (1983), slip at the measuring surfaces in rotationa
viscometers must be avoided. This was achieved to some exten
by sticking abrasive paper to the parallel plate system and
initially by sticking sand grains to the surfaces of the APC
system and subsequently, by engraving the surfaces. The
asperity size in each case was about 100 microns.

Table 2.   Composition of Oilwell cement (%)

Oxides									
SiO_2	Al_2O_3	Fe_2O_3	CaO	MgO	SO_3	K_2O	Na_2O	LoI	I.R
21.78	3.78	4.99	64.32	0.74	2.34	0.38	0.09	1.01	0.1

Compounds				
C_3S	C_2S	C_4AF	C_3A	$C\bar{S}$
57.6	19.0	15.2	1.39	3.98

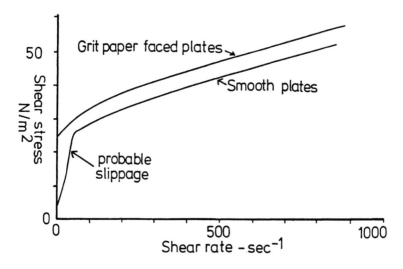

Fig.3. Effect of surface finish on flow curve.

3 Experimental Programme

3.1 Materials and mixes

Class G oilwell cement (Blue Circle Industries PLC) with the
composition given in Table 2 was used in all tests. It had a
thickening time of 110 minutes (API Schedule 5). All slurries
used deionised water.

3.2 Mixing Methods

Preliminary experiments used slurries prepared according to API
specification 10 (API, 1986) using a Waring blender. However,
the large (600ml) sample so prepared was not homogeneous and it
was difficult to obtain sufficiently representative samples for
testing. Furthermore, the blender does not control the shear
rate and hence the energy input during mixing with sufficient
constancy for studying its effect on yield stress. Therefore
the APC geometry in the rheometer was used to preshear a sample
of slurry which had been mixed by hand according to the
following schedule. Cement and water were blended by hand in a
beaker with a small stainless steel spatula for 60 sec. 16g of
slurry was poured into the centre of the annular plate and
transferred to the rheometer. The plate was raised to the
measuring position and the rheometer started to move at 3
minutes after first contact of water and cement. The mixing
phase is preprogrammed and at the end the rheometer stops and
the slurry sample is stationary for a preprogrammed time. Then
the testing phase starts. The whole process is completed
without any handling of the sample and so the shear rates and
stresses are recorded at all times. This mixing and testing
procedure required the standard flow curve software to be
modified to allow the peak stress to be held and the sample to

stand before starting the measuring stage.

3.3 Test programme
The complete test programme provided information on the effects
of mixing energy, time and slurry water content on the yield
stress as measured in the three geometries and under different
conditions of applied stress ramp rate, but only selected
features will be described here. All experiments were at 25°C.

4 Results and Preliminary Discussion

4.1 Comparison of different geometries and surfaces
Fig. 3 shows flow curves obtained for slurries of 0.485
water/cement ratio mixed in the Waring blender according to API
specification 10, and tested in the parallel plate geometry,
(i) with smooth stainless steel surfaces and (ii) with plates
to which grit paper had been stuck.
   The mean yield stress (defined as the stress at which flow
first occurs) of four samples was 2.24 N/m^2 with smooth plates
and 25.7 N/m^2 with grit surfaced plates. The spread of values
was over 30% in the former case and under 10% in the latter.
This suggests that slippage is reduced by roughening the
surfaces but not necessarily overcome completely. When flow is
established, the plastic viscosity, as indicated by the slope of
the lines, is unaffected by the nature of the surface,
confirming Mannheimer's finding that slippage has the greatest
effect at stresses around the yield stress and suggests that the
lubricating layer is remixed into the slurry by the motion of
the plate as the flow curve is built up. Table 3 shows yield
stresses of repeat samples of nominally identical slurry
determined in all three geometries. The rate of increase of
torque was different in each case so as to give a constant

Table 3. Comparison of yield values in the different geometries

Measuring system	Surface	Yield stress N/m^2
Parallel plate	smooth	3.05, 3.05, 3.05, 2.13, 3.05
	grit paper	21.0, 27.2, 29.0, 22.9, 25.9
Vane in cup	-	38.7, 42.7, 33.5, 35.7, 30.8
A.P.C.	smooth	25.9, 23.8, 20.8
	sand faced	35.9, 53.9, 43.9, 51.8

applied stress ramp rate (61 N/m^2 per minute) according to the formulae given in Table 1.

These results confirm the superiority of the roughened APC geometry, even over the vane, in overcoming slippage, although it has to be noted that the variation between repeat samples is large and the difference may be within experimental error. That the vane does not give the expected highest yield stress may be due to the non-ideal immersion geometry required by the dimensions of the rheometer. The results give added support to the selection of the APC for the main programme.

4.2 Calculation of mixing energy

The rate of energy dissipation per unit volume in viscous flow is:

$$E \quad = \quad \eta \dot{\gamma}^2 = \tau \dot{\gamma}$$

so the total energy per unit mass in time t_s is:

$$E_T/M \quad = \quad E\,t\,V/\rho V = \tau/\rho \int_o^{t_s} \dot{\gamma}\,dt$$

where ρ is the density of.the slurry. In the CS rheometer the shear stress τ is constant, governed by the programmed torque via the appropriate conversion factor in Table 1. Therefore, the total energy input during the mixing stage in the APC geometry is proportional to the area under the curve of shear rate against time. As an example of the total energy input, a slurry of density 1900 kg/m^3 (0.44 w/c) was sheared in the APC at constant stress equivalent to a torque of 25000 dyne cm. The average angular velocity was 21 radians/sec from which E_T/M is 0.45 kJ/kg. This is the same order of magnitude as the so called specific mixing energies (SME) calculated by Orban et al (1986) for slurries prepared by various industrial methods. The standard API laboratory procedure was found to give SME = 1.0. Selection of conditions enabled a range of shearing energies to be studied (Table 4).

Table 4. Mixing times for various specific mixing energies

Energy input kJ/kg	0.55	1.38	2.75	4.13	5.5
S M E	0.1	0.25	0.5	0.75	1.0
Mixing time (min)	1.15	2.9	5.75	8.6	11.5

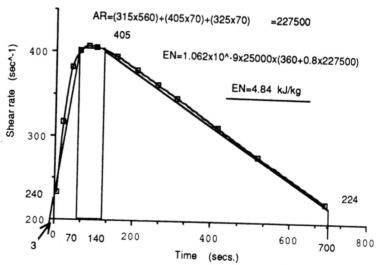

Al=(3x240)/2=360

AR=(315x560)+(405x70)+(325x70)   =227500

405

EN=1.062x10^-9x25000x(360+0.8x227500)

EN=4.84 kJ/kg

Fig.4. Area under a typical shearing energy curve.

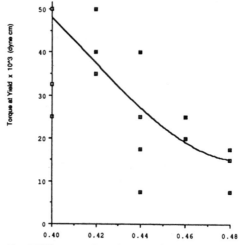

Fig.5. Effect of w/c ratio on yield stress

In practice, a desired mixing energy cannot be preprogrammed reliably and the actual energy can be calculated only from the observed variation in shear rate with time. Fig. 4. shows a typical curve at constant torque of 25000 dyne cm and method o calculating the input. The form of the curve in Fig. 4 is entirely consistent with the behaviour which is well established in the testing of cement pastes in controlled shea rate viscometers (Tattersall & Banfill 1983).   The first stage with shear rate increasing steeply corresponds to the structur breakdown; the peak is where all the structure has been broke

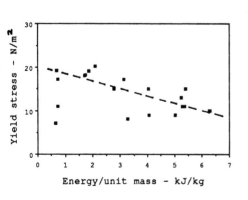

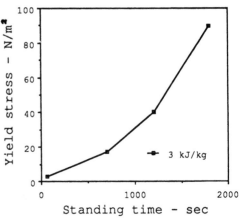

3 kJ/kg

Fig.6 Effect of mixing energy/ unit mass on yield stress.

Fig.7 Effect of standing time on yield stress.

down; and the third stage is where the slurry thickens again as the effects of early hydration reactions start to dominate. The time to reach the peak shear rate (broken down structure) was consistently 60 - 100 sec. Fig. 5 shows the effect of w/c ratio on the yield stress determined with the APC geometry. These yield stresses have to be taken into account when selecting the peak torque to be programmed in the mixing shear cycle, as it is necessary to exceed the yield stress by a sufficient margin in order to control the shear in the sample.

4.3 Effect of mixing energy on yield stress

Fig. 6 shows that as mixing energy, calculated as in the previous section, increases, the yield stress of a 0.44 w/c ratio slurry decreases. The mixing energy may increase as a result of prolonged mixing times but in slurries of practical importance the time until testing or use may include a standing time during which the slurry is undisturbed. Fig. 7 shows that the yield stress/mixing energy ratio increases with increasing standing time, after completion of the mixing stage. Curve fitting on the experimental data gives the following formula for the variation of yield stress with mixing time and standing time.

$$\tau_y = \rho \frac{E_T}{M} e^{(0.1724(t_{stand})^{\frac{1}{2}} - 11.1607)} t_{mix}^{(-0.2619 - 0.1284 \ln t_{stand})}$$

Fig. 8 shows that the agreement between experimental points and the best fit curves is acceptable, but it must be recognised that this empirical relationship applies only within the range of these experiments.

133

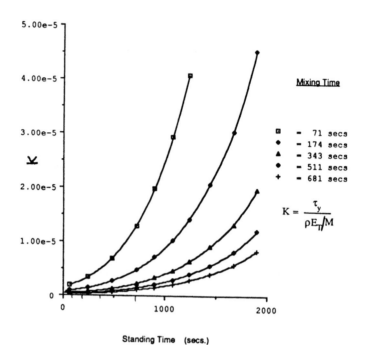

Fig.8 Relationship between mixing, standing and yield stres

5 General discussion

5.1 Experimental considerations

These results show that the Carrimed CS rheometer may be used
measure the yield stress of oil well slurries and that the
control system and software is sufficiently flexible to allow
the effect of shear history to be studied. However, various
experimental difficulties had to be overcome, the most
inconvenient of which was a tendency of the rheometer to shut
down due to mains-borne electrical interference. When this
occurred the control system would abort the run, shut down th
rheometer and lower the sample away from the measuring system
The longer the run in progress (e.g. preshearing and standing
the greater the likelihood of shutdown before the results cou
be obtained. Careful electrical isolation of the system
eventually reduced the frequency of shutdown. Secondly, the
geometry is very sensitive to sample size and preparation. T
gap must be exactly filled and quite small overflows or
underfilling can affect the measured yield stress, although
corrections can be made for this (Kitching, 1989).

5.2 Implications for practice in the oil industry

As already noted, the pumping pressures needed to circulate
cement slurry to effectively remove residual mud and spacer
fluids depend on its yield stress, so reliable measurement is
important. This work has shown again that slippage has a
significant effect on the measured yield stress and suggests

that the standard method of yield stress determination in the oil industry, based on a Fann viscometer with smooth concentric cylinders, cannot be expected to give a reliable measurement. Secondly, the yield stress, defined as the stress at which continuous flow first occurs, can only be measured in a controlled stress rheometer. Extrapolation back to zero shear rate in the Fann procedure is misleading.

The yield stress depends on the mixing energy input to the slurry. The mixing energy put in by the Waring blender in the API procedure (5.5 kJ/kg as estimated by Orban et al) is about 10 times that put in by field mixing equipment, but the blender is insufficiently controllable to allow these lower energy input levels to be reproducibly achieved. However, the mixing shear programme developed for the controlled stress rheometer is capable of achieving them. It follows that the yield stresses determined in the laboratory and routinely used for predicting the behaviour of oil well cement slurries may be a considerable underestimate of the yield stresses achieved in practice. The measured yield stress may be one tenth of the true yield stress due to slippage, while the yield stress of a slurry produced in the field may be ten times that prepared according to the API standard procedure, particularly if the slurry remains stationary for any length of time.

6 Conclusions

At ambient temperature and pressure, yield stress decreases with increasing water content and increasing mixing energy but increases with increased standing time.
Controlled stress rheometry may be a better way of testing slurries for which better precision is needed and where the previous shear history is important and can be quantified.

7 References

API (1986) **Specification for materials and testing for well cements,** American Petroleum Institute, Dallas (API specification 10, 3rd edition).
Cross, M.M., Kaye, A. (1986) Techniques for the viscometry of suspensions, **Polymer Engineering Science,** 256, 121–126.
Cross, M.M., Kaye, A. (1987) Simple procedures for obtaining viscosity/shear rate data from a parallel disc viscometer, **Polymer,** 28, 435–440.
Kitching, D.R. (1989) **Yield stress of an oilwell cement slurry using a controlled stress rheometer,** Ph.D. Thesis, University of Liverpool, U.K.
Nguyen, Q.D. (1983) **Rheology of concentrated barite residue and suspensions,** Ph.D. Thesis, Monash University, Australia.
Mannheimer, R. J. (1983) Effect of slip on flow properties of cement slurries can flaw resistance calculations. **Oil & Gas**

Journal, Dec 5, 144-147.

Orban, J. A., Parcevaux, P. A., Guillot, D. J. (1986) Specific mixing energy: a key factor for cement slurry quality. SPE 15578, in **61st Annual Conference of the Society of Petroleum Engineers,** New Orleans 1-5.

Tattersall, G. H., Banfill, P. F. G. (1983), **The Rheology of Fresh Concrete,** Pitman, London.

8 Acknowledgements

We are grateful for the financial support of the Science and Engineering Research Council and B.P. Research Centre Ltd through a CASE studentship.

15 THE USE OF THE SHEAR VANE TO MEASURE THE GEL STRENGTH AND DYNAMIC YIELD STRENGTH OF OILWELL CEMENT SLURRIES AT HIGH TEMPERATURE AND PRESSURE

J. KEATING and D.J. HANNANT
Civil Engineering Department, University of Surrey, UK

Abstract
A parametric survey of typical pumping conditions in North Sea well cementing has shown that although the flow of the cement slurry may occasionally be turbulent in the annulus, the flow is generally laminar in the pipe with a plug diameter in excess of 70% of the pipe diameter. This paper examines the properties of the material in the plug obtained by measuring the 'gel strength', defined as the static yield strength at zero shear rate, by the use of a six bladed shear vane. This technique has the advantage of avoiding high water content slip layers such as may form at the surface of smooth bobs. The development of the shear vane technique into the high pressure and high temperature regime by the use of a modified pressurized consistometer is described. Data are presented showing the effect of pressures up to 36 MPa and temperatures up to 52°C on the gel strength and dynamic yield strength of two cement slurries used in the oil industry. It is shown that gel strengths in excess of 150 Pa are measured by the vane test after forty minutes of heating and values up to 2000 Pa are observed after seventy minutes heating. The effect of pressure alone implies that the slurries may not behave as fully saturated undrained systems.
Keywords Shear vane, oil well cements, gel strength, yield strength.

1 Introduction

Oil wells are drilled to depths of several kilometres where the temperature can reach 200°C with pressures up to 140 MPa. Oil well cement slurries are used to seal the gap between the rock wall and the well casing and also to seal the annulus between successive lengths of casing with different diameters. Important parameters for the prediction of the pressures in the cementing process are the gel strength and dynamic yield strength of the slurry under down-hole conditions. As there is continuing controversy regarding the exact meaning of these two strengths and the correct measurement technique the definitions used in this paper are explained in section 2.

2 Gel strength and dynamic yield strength

In high solids content suspensions such as cement slurries with strong particle interactions, the bulk of the suspension may not only have a static yield strength at zero shear rate, denoted here as 'gel strength', but also due to the nature of the slurry, may have a dynamic yield strength lower than the gel strength as the slurry breaks down under shearing.

Thus, under the application of a small stress, the system will deform elastically with finite rigidity, but when the stress exceeds the gel strength a continuously increasing deformation will occur at a lower shear stress which is denoted 'dynamic yield strength', this being a dynamic rather than static property of the suspension. The gel strength therefore can be regarded as a property denoting a transition between solid-like and liquid-like behaviour. However, at slow rates of shearing, the rate of chemical build-up of structure may equal the rate of breakdown due to shearing so that the gel strength and dynamic yield strength may be similar. Keating and Hannant (1989a).

The shear vane was chosen to make the measurements because it has been established by Mannheimer (1982), Bannister and Benge (1981), Dzuy, and Boger (1983) and (1985), and Haimoni and Hannant (1988) that a high water content slip layer may be developed at the smooth bob surface in traditional coaxial viscometers such as the Fann viscometer. The torque value obtained will then relate more to the viscosity of the suspending medium than that of the suspension. According to Haimoni and Hannant (1988) this yield value may be less than 50% of the gel strength obtained from shearing the bulk material with a vane. Mannheimer (1982) also states that the presence of the slip layer can cause laboratory data to grossly underestimate the flow resistance of cement slurries in the annulus.

3 Flow regimes in cementing oil well casing

In order to provide a correct conditioning procedure for oil well cement slurries before vane tests are carried out, it is necessary to determine the types of flow which may occur during the field cementing of oil well casings. This requirement is emphasized by Orban et al (1986) who state that the mixing and conditioning shear actions should account for "pumping the cement slurry into the casing and displacing it into the casing to formation annulus". It is essential therefore to know whether the flow regime in the slurry is turbulent, mostly laminar or mostly plug. It is inevitable that some plug flow will be involved if turbulence is not achieved because, for a Bingham type material with a yield point, the local shearing stress in the pipe near the axis will be less than the yield value and hence the material within this radius will move as a solid plug. In the plug which is unsheared, the gel strength may build-up relatively rapidly and this gelled material will need to be broken down when passing through the shoe. It will also be sheared in the irregular annulus between the production casing and the rock and also in the eccentric casing-casing annulus.

A range of pipe sizes, pumping rates and slurry parameters were therefore chosen covering typical North Sea conditions and a parametric analysis was carried out for the following ranges of values to establish the flow regimes which could occur in practice.

3.1 Parameters typical of North Sea conditions

Pumping Rates   0.0053 - 0.032m^3/S (2-12 bbl/min)

Hole   a)   0.311 m. dia hole with 0.245 m O.D. (0.224 m I.D.) casing
casing   (12.25 in dia. hole with 9.63 in O.D. (8.84 in I.D.) casing
sizes   b)   0.216 m dia. hole with 0.178 m O.D.(0.162 m I.D.) casing
                 (8.5 in dia. hole with 7 in O.D. (6.37 in I.D.) casing

Casing lengths   1,500 - 2,500 m (4,920 - 8,200 ft)

Slurry   Plastic viscosity 0.01 - 0.16 Pa.s (10 - 160 c. poises)
                  (0.00675-0.108 lb mass
                                  ──────
                                   ft s
Yield Stress   10 - 1,000Pa (0.209 - 20.9 lb f/ft^2)
Density   1,580 and 1,900 kg/m^3 (13.2 - 15.8 lb/US gal)

3.2 Prediction of flow parameters

The procedures for a Bingham Plastic fluid defined in American Petroleum Institute Specification 10 (1986) were used to calculate the Hedstrom number, Reynolds number and minimum rate for turbulent flow. In addition, the plug diameter in laminar flow conditions in a pipe was calculated using the Buckingham equation and analytical procedures detailed in Wilkinson (1960). Computer programs were set up to solve the equations and tables were generated for the complete range of variables.

For pipe flow, turbulent flow was only predicted for the pipe of 0.162 m inside diameter with a slurry yield stress of 10 Pa, and pumping rates in excess of 0.030 m^3/s. In all other cases laminar flow was indicated. In the annulus, the equations require the assumption that the annulus is symmetrical but it is almost inevitable that some eccentricity is present. In the narrowest portion of the annulus, the drilling mud may be very difficult to displace during the cementing operation and the velocity in this region may be very slow. In the larger section of the annulus, flow may be approaching turbulence. Thus at any given section the velocity profile of flow will be variable and unknown.

The plug diameters calculated using the Buckingham equation for flow conditions shown to be laminar are smallest when the slurry yield stress is lowest, the slurry plastic viscosity is highest and the pumping rate is highest. The slurry density was not a parameter in these calculations and the different pipe lengths i.e. 1,000 m and 2,500 m had negligible effect.

An example of the results from these calculations is shown in Table 1 for a pipe of 0.179 m outside diameter and 0.162 m inside diameter. The plug size was generally in excess of 50% of the pipe internal diameter and in most cases was in excess of 80% of the pipe diameter.

Table 1 Theoretical ratio of plug diameter to pipe internal diameter
Pipe outside diameter 0.178 (m)
Pipe inside diameter 0.162 m   Pipe length 2500m   Slurry density 1580 kg/m³

Plast Visc. Pa.s	Flow m³/s										
		0.0053			0.0012			0.0022			0.032
0.01	yield (Pa)	10	0.921	yield (Pa)	10	0.882	yield (Pa)	10	0.843	yield (Pa)	10	turbul
		100	0.975		100	0.962		100	0.949		100	0.939
		1000	0.992		1000	0.988		1000	0.984		1000	0.980
0.06	yield (Pa)	10	0.812	yield (Pa)	10	0.724	yield (Pa)	10	0.640	yield (Pa)	10	turbul
		100	0.939		100	0.909		100	0.877		100	0.852
		1000	0.980		1000	0.971		1000	0.960		1000	0.952
0.11	yield (Pa)	10	0.750	yield (Pa)	10	0.639	yield (Pa)	10	0.536	yield (Pa)	10	turbul
		100	0.917		100	0.877		100	0.835		100	0.803
		1000	0.974		1000	0.961		1000	0.946		1000	0.936
0.16	yield (Pa)	10	0.703	yield (Pa)	10	0.578	yield (Pa)	10	0.465	yield (Pa)	10	turbul
		100	0.901		100	0.853		100	0.803		100	0.765
		1000	0.968		1000	0.852		1000	0.936		1000	0.923

3.3 Effect of the analysis of the flow regimes on the test programme

It is clear that for a substantial part of the pumping operation a considerable portion of all slurries are effectively unsheared in the form of a plug. This is particularly the case in the pipe as opposed to the annulus. The gel strength may therefore build up rapidly in these conditions and hence a substantial part of the test programme was devoted to shear vane tests on stationary rather than continuously agitated slurries.

4   Test equipment

4.1 Shear vane

In order to overcome the problem of high water content slip layers with smooth bobs we have developed the use of a six bladed vane, as used in measurements on soft clays. This technique has the great advantage that the slurry shears within itself and there is no slip layer before shearing to affect the results The test has been described in detail by Haimoni and Hannant (1988) and Keating and Hannant (1989a). The vane used in this work had the dimensions 33 mm diameter by 61 mm high. This vane diameter ensured that shear occurred at the tips of the vane and not at the interface between the slurry and the pot.

4.2 Modification of pressurised consistometer

The major item of equipment was a Chandler Model 7 pressurized consistometer shown in Figure 1, of a type commonly used in the oil industry to measure thickening times. This equipment is capable of applying pressures of 200 MPa at temperatures up to 200°C. The vane replaced the stirring paddle without any alteration to the slurry pot or to the torque

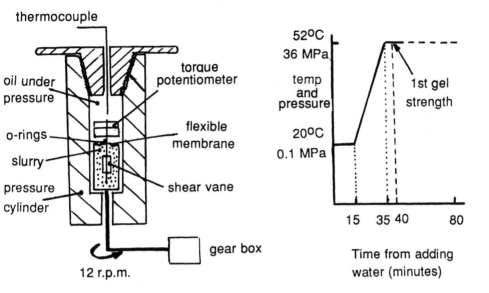

Figure 1. Main features of pressurised consistòmeter.

Figure 2. Slurry conditioning schedule.

measuring device. The standard rotation speed of this equipment is 150 rev/min but this was too fast for gel strength measurements to be recorded. The main modification therefore was the addition of an auxiliary motor and gear box coupled to the main drive shaft to rotate it at 12 rev/min. For most slurries, the time to peak torque at this rotation speed was more than one second which was well above the minimum time of 400 ms recommended by Keating and Hannant (1989a). For recording the torque-time curve, the output from the torque head was fed into a Y-t recorder.

Because some of the slurries were expected to have relatively low gel strengths at early ages, it was necessary to check that O-ring friction was not a significant part of the torque reading. A series of rotational tests was therefore carried out at 12 rev/min with water in the pot of the pressurized consistometer and pressures of 0.7 MPa, 34 MPa and 69 MPa were applied. It was found that no torque reading was registered which implied that O-ring friction was less than that in the torque measuring head.

5 Test programme

5.1 Mix proportions
Two slurries used in the oil industry were used for the shear vane tests the mix proportions being shown in Table 2.

141

Table 2 Mix proportions by weight

	Proportions (% by weight of cement)	
Material	Neat Class G	Low Density
Class G cement	100	100
Pre-blended Wyoming bentonite	-	8
Antifoam agent	0.1	-
Water (distilled)	44	90
Slurry density (kg/m^3)	1900	1580

5.2 Slurry conditioning

It is well known that the rheological properties of cement slurries are dependent on the shear history before the start of the test and because of the relevence of the data to the drilling industry the mixing procedure was that specified by the American Petroleum Institute (1986). The AP I also specifies typical rates of temperature rise and pressure increase for testing oil well cements. Based on Schedule 5 of the specification the slurry conditioning schedule shown in Figure 2 was followed.

The time required to mix the slurry, place it in the pressurized consistometer and re-assemble the equipment occupied about fifteen minutes which was longer than the API specification due to additional problems with aligning the shear vane with the torque measuring head. Heating then started with the slurry kept stationary to simulate the material in the plug until the auxillary gear box drive was switched on, the earliest readings being taken at about forty minutes from the addition of water or twenty five minutes after the start of heating. In order to avoid reshearing the same material, each test at a different time was on a separate mix and a range of times was used up to a maximum of about seventy minutes from the start of heating. In order to investigate the effect of pressure on gel strength, an identical test regime was followed except that the pressure was reduced from 36 MPa to one atmosphere over a period of five minutes before the slurry container was rotated. A total of 34 tests were made on the neat Class G mix and 16 tests on the low density mix.

6 Results and Discussion

6.1 Tests at 52°C and 36 MPa

Both the neat Class G and the low density slurries showed similar shapes of shear stress-time curves after 25 minutes and 65 minutes from the start of heating. Figure 3 shows curves at these two times for the Class G slurry and it can be seen that the earlier readings generally lack the peak which is shown

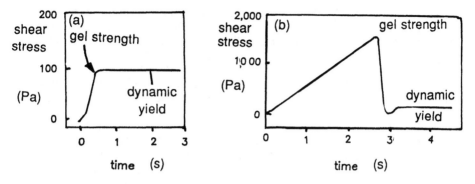

Figure 3. Shear stress-time curve at 52ºC and 36 MPa.
(a) 25 minutes after start of heating (b) 65 minutes after start of heating

at later ages. Both curves are characterised by an initially linear portion but
after the peak torque was reached in Figure 3 (b) the resultant sudden release
of energy in the spring combined with the reduction in shear strength on the
slip surface caused the nearly vertical trace on the Y-t recorder which
overshot the zero mark before returning to some residual shear strength
torque value at about 7% of the peak torque. For Figure 3 (a) the gel strength
and dynamic yield strength had similar values The relations between gel
strength and time for the two slurries is shown in Figure 4. It was surprising to
find that the curves for the two materials were similar even although the low
density slurry had twice the water content of the neat Class G slurry. The

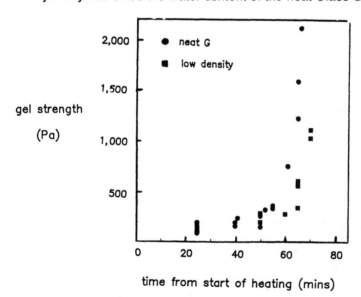

time from start of heating (mins)

Figure 4. Relation between gel strength and time.

materials are characterised by the very rapid gain of gel strength after sixty minutes of heating with strengths in excess of 2,000 Pa being reached after 70 minutes of heating. Even at heating times of less than 35 minutes the gel strength had exceeded 150 Pa, which is a level which is considered difficult to pump by the oil industry. This is much sooner than normally expected using traditional smooth bob techniques and therefore some re-thinking of pumping parameters may be necessary if shear vane technology becomes accepted as a valid measurement technique.

The relationship between dynamic yield strength and time from the start of heating for the two slurries is shown in Figure 5. These stresses do not show the rapid increase after 50 minutes of heating which is a feature of gel strength development. Instead there is a steady increase between about 30 minutes and 65 minutes, the maximum dynamic yield strengths being less than 300 Pa.

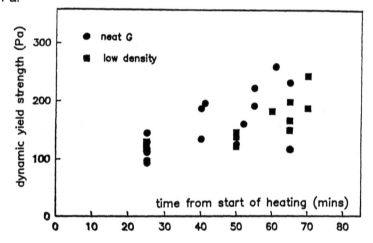

Figure 5   Relation between dynamic yield strength and time

6.2   Theoretical effect of pressure on gel strength

In a mixture of water and cement, shear stress can only be resisted by the skeleton of solid particles by means of forces developed at interparticle contacts. At the beginning of the tests, where water forms a continuous phase, the cement slurry situation may be similar to a fully saturated undrained triaxial test on clay. In this case, an increase in pressure will increase the pore water pressure but it will not cause the particles to be pressed closer together and hence no change in shear strength would be expected with increase in pressure. This is because the effective stress transmitted through the cement skeleton will not change.

However, this hypothesis assumes that the solids and water are incompressible, there is full saturation, and that there is no air in the system. In practice, some air will almost certainly be present and the water is continuously being removed from the system by chemical reaction leading to chemical shrinkage, Keating et al (1986 b). The rigidity of the cement skeleton will also increase with time. These effects may allow the solid particles to move closer together in the consistometer pot as the oil pressure is increased and if this is the case, an increase in shear strength measured by the vane would be expected. In practice, an increase in effective stress and hence shear strength could occur as well depth increases which could have practical implications.

6.3   Results from different oil pressures

The results for gel strength and dynamic yield strength for neat Class G slurry are shown in Figure 6 on an enlarged scale. The effect of pressure reductions at twenty five minutes and forty minutes is to reduce the gel strength and dynamic yield strength implying a reduction of effective stress on the particles. This assumes that the removal of pressure did not break down the weak structures in the gel. However, after the setting reaction accelerated from sixty minutes onwards it was not possible to judge whether pressure had an effect on gel strength because very small increments of time had such a large effect on gel strength that the effect of pressure was swamped by minor time differences.

Similar tests carried out on the low density slurry did not show such marked effects of pressure although insufficient data was obtained to be confident of the effects of pressure. However, higher water content slurries might allow the pore water pressure to increase without causing an increase in effective stress on the particles.

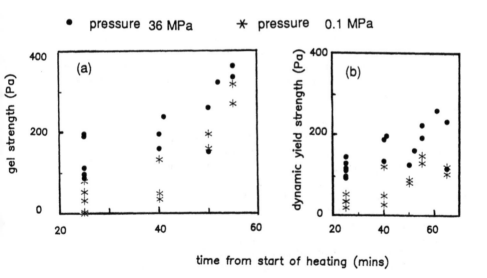

Figure 6   Effect of 36MPa pressure on gel strength and dynamic yield strength
for neat Class G slurry
(a) Gel strength   (b) Dynamic yield strength

7   Conclusions

A parametric study has indicated that when cement slurries are pumped during oil well cementing, a substantial plug is likely to exist before the cement is displaced into the annulus.

The standard type of pressurized consistometer used in the oil industry can be successfully modified to use the shear vane to measure gel strengths which are considerably greater than those measured using smooth bobs.

Pressures of 36 MPa can have an effect on the gel strength and dynamic yield strength of some slurries.

Acknowledgments

The authors gratefully acknowledge the financial support provided by the Marine Technology Directorate Ltd of the Science and Engineering Research Council. Also, the project would not have been possible without the pressurised consistometer funds which were provided by B.P. Petroleum Development Ltd.

References

American Petroleum Institute (1986). A.P.I. Specificiation 10. **Specification for materials and testing for well cements.** 3rd Edition.

Bannister, C.E. and Benge, O.G. (1981) Pipe flow rheometry: Rheological analysis of a turbulent flow system used for cement placement. **Soc. Petr. Eng. SPE 10216.**

Dzuy, N.Q. and Boger, D.V. (1983) Yield stress measurement for concentrated suspensions. **J. Rheology.** 27 (4) 321 - 349.

Dzuy, N.Q.and Boger, D.V. (1985) Direct yield stress measurement using vane method. **J. Rheology** 29 (3) 335 - 347.

Haimoni, A. and Hannant, D.J. (1988). Developments in the shear vane test to measure the gel strength of oil well cement slurry. **Advances in Cement Res.**1 (4) 221 - 229.

Keating, J. and Hannant, D.J. (1989a). The effect of rotation rate of gel strength and dynamic yield strength of thixotropic oil well cements measured using a shear vane. **J. of Rheology,** 33 (7) 1011 - 1020.

Keating, J., Hannant, D.J. and Hibbert, A.P. (1989b). Correlation between cube strength, ultrasonic pulse velocity and volume change for oil well cement slurries. **Cem. and Con. Res.** 19. 715 - 726.

Keentock, M., Millthorpe. J.F. and O'Donovan, E. (1985) On the shearing zone around rotating vanes in plastic liquids: Theory and equipment. **J. Non Newtonian Fluid Mech.** 17, 23 - 25.

Mannheimer, R.J. (1982) Rheological evaluation of cement slurries. **American Petroleum Institute.** Final Report No. SWR-6836.

Orban, J., Parcevaux, P. and Guillot, D. (1986) Influence of shear history on rheological properties of oil well cement slurries. **8th International Congress on the Chemistry of Cement.** Rio de Janiero, Brazil, 1, 243 - 247.

Wilkinson, W.L. (1960). **Non-Newtonian fluids. Fluid Mechanics, mixing and heat transfer.** Pergamon Press.

16 LAMINAR AND TURBULENT FLOW OF CEMENT SLURRIES IN LARGE DIAMETER PIPE – A COMPARISON WITH LABORATORY VISCOMETERS

R.J. MANNHEIMER
Southwest Research Institute, San Antonio, Texas

Abstract
This paper describes methods for measuring wall slip that can be applied to chemically reactive materials such as cements, and compares the rheological properties of a cement slurry measured with different laboratory viscometers that required large corrections for wall slip. Experimental results pertaining to laminar, transitional and turbulent flows of cement and clay slurries in a large diameter pipe are discussed in terms of theoretical predictions. A comparison is also made of rheological properties of cement slurries measured with a laboratory viscometer and a large-scale pipe viscometer.
Keywords: Cement Rheology, Wall Slip, Laminar-Turbulent Flow

1 Introduction

Many factors can contribute to the problem of obtaining reliable measurements of the viscosities of cement slurries with laboratory viscometers. In particular, once water is added to dry cement, hydration and chemical reactions start to build up a sensitive structure that can be broken down by shear. Consequently, rheological measurements of cement slurries usually depend on both the shear and time histories of the sample. Furthermore, although many cement slurries appear to be stable while they are at rest, the shearing action of a viscometer can accelerate sedimentation and produce anomalous rheological data (Bhatty and Banfill, 1982). Even when these factors have been properly controlled or accounted for, the rheological properties of cement slurries can still be difficult to measure with laboratory viscometers because of wall slip (Bannister, 1980; Mannheimer, 1983).

While the exact cause of wall slip can be quite complex, it can usually be explained in terms of a thin film of liquid (often less than one μm) that lubricates the walls of the viscometer. This simple explanation is illustrated in Fig. 1. In this case, the walls of a concentric cylinder or Couette viscometer with a narrow shear gap have been simulated by two flat plates. However, the same explanation could be used to explain wall slip in tube viscometers. The effect of wall slip is particularly troublesome for cement slurries that have a yield value in that it provides a mechanism for flow at shear stresses below the yield value. For the example shown in Fig. 1, the velocity of the cement slurry (V) does not change across the shear gap (H). Instead, all of the shear takes place in the thin liquid film (E << H), which is referred to as 100% slip. While no meaningful rheological information is provided at 100% slip, this condition is often encountered with cement slurries. A common feature of wall slip is that the measured viscosity will be lower for smaller shear gaps. Thus, errors in the viscosity are magnified by the small shear gaps (typically 1 mm) of a laboratory viscometer and could seriously underestimate the pressure losses in large diameter pipes where slip should be negligible.

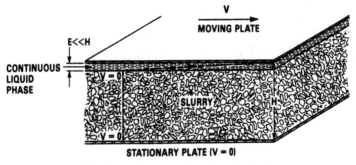

Fig. 1.   A simple explanation of wall slip

In order to compare rheological measurements of cement slurries made with different laboratory viscometers, one must first determine the relative importance of slip and, if necessary, correct the data for slip. However, to the writer's knowledge, no comparison of the rheological
properties of cement slurries measured with different laboratory viscometers has ever been
made in which the potential effects of slip were considered. Consequently, this paper will
explain how to detect wall slip with laboratory viscometers, and will compare rheological measurements of cement slurries measured with Couette and tube viscometers in which relatively
large corrections for slip were necessary.

In addition to the problem of predicting friction factors for the laminar flow of cements in
large-diameter pipes from laboratory rheological data, oil-field cementing operations often
involve turbulent flows. For example, the displacement of drilling mud by cement and the
removal of mud cake from the formation (both of which are necessary for good bonding of the
cement) are usually more effective if the flow is turbulent. However, very little is known about
the requirements for the transition from laminar to turbulent flow with highly non-Newtonian
cement slurries. Since cement slurries are more non-Newtonian at low shear rates, experiments
to study laminar-turbulent transition must be conducted in large-diameter pipes in order to produce large inertial effects at relatively low shear rates. Therefore, this paper will discuss experiments dealing with laminar and turbulent flow of cement slurries in large-diameter pipe and
compare these results with theoretical predictions and laboratory data.

2   Comparison of rheological measurements made with Couette and tube viscometers

2.1 Analysis of wall slip
Mooney (1931) devised an ingenious method for analyzing Couette viscometer data for wall
slip that requires measurements of the torque (T) as a function of the rotational speed (Ω) for
three different bob-to-cup ratios $(\beta = R_b/R_c)$. In this same paper, Mooney also discussed a
method for analyzing tube viscometer data that requires measurements of flow rates (Q) and
pressure gradients $(\Delta P/L)$ with at least three different size tubes. Unfortunately, the necessity of
conducting triplicate experiments is not well suited for chemically reactive materials, like
cement slurries, whose viscosities tend to increase with time. The most commonly used method
for detecting slip is to compare measurements obtained with smooth and grooved or roughened
cylinders (Cheng, 1984). Since rough surfaces tend to reduce slip, apparent viscosities will
usually be higher with roughened cylinders when slip is an important factor. The problem with

this method is that it does not ensure that the roughened cylinders have completely eliminated slip, nor does it provide the information that is needed to correct data for slip.

The analysis of wall slip for Couette viscometers using only two shear gaps $H = (R_c - R_b)$ is hampered by the fact that there is no simple analogy to the consistency parameters of wall-shear stress (τ_w) and Newtonian shear rate $(4V/R)$ that can be used to measure wall slip with tube viscometers. In particular, provided that kinetic energy and entrance effects are negligible, time independent materials should show no dependency on the inside radius of the tube (R), when the measured pressures and flow rates are expressed in terms of τ_w and $4V/R$. Moreover, the slip velocity (V_s) can be calculated from measured differences in the Newtonian shear rates at the same shear stress with two different size tubes (Fredrickson, 1964):

$$V_s = \frac{(4V/R)_1 - (4V/R)_2}{4(1/R_1 - 1/R_2)}, \; \tau_w = c \tag{1}$$

and the corrected Newtonian shear rate at this shear stress is

$$\left(\frac{4V}{R}\right)^* = \frac{4V}{R} - \frac{4V_s}{R} \tag{2}$$

For narrow gap concentric cylinder viscometers $(0.9 \le \beta < 1)$, and for materials that can be characterized by a restricted range of values for the flow index $(0.1 \le n' \le 1)$, a similar procedure can be used by expressing measurements of the torque and rotational speed in terms of a mean shear stress (τ_m) and a mean shear rate (V/H) (Mannheimer, 1982, 1983, 1985 and 1988):

$$\tau_m = \frac{T}{2\pi R_m^2 L} \tag{3}$$

$$\frac{V}{H} = \frac{R_m \Omega}{R_c - R_b} \tag{4}$$

$$R_m = \frac{R_b + R_c}{2} \tag{5}$$

The fact that τ_m and V/H can serve as consistency parameters, which can be used to measure wall slip, is illustrated in Fig. 2. The solid curve in Fig. 2 represents a Bingham Plastic $(\tau - \tau_o = \mu_{pl}\dot{\gamma})$ with $\tau_o = 10$ Pa and $\mu_{pl} = 10$ mPa s. On the other hand, the values of τ_m and V/H in Fig. 2 were calculated from the Reiner-Rivlin equation for the same Bingham Plastic in a Couette viscometer for two values of β, which for a fixed value of R_c is equivalent to two different shear gaps $[H = R_c(1 - \beta)]$.

The results in Fig. 2 show that τ_m and V/H are excellent approximations of Bingham Plastic flow, and that there is no significant difference between values of V/H for the same value of τ_m as long as τ_m is not too close to τ_o. Since τ_o is not generally known beforehand, this problem can be avoided by limiting the use of τ_m and V/H to data where the slope of the flow curve is

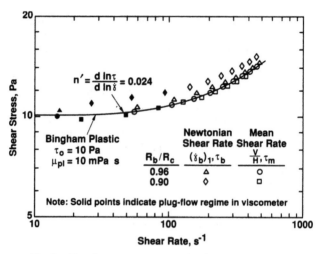

Fig. 2. Consistency parameters for Couette viscometer

not too close to zero (i.e., $0.1 \leq n' \leq 1$). For example, at $\tau_m = 1.04\ \tau_o$ in Fig. 2, $n' = 0.024$; consequently, this point would not be included in the analysis of Couette viscometer data.
Yoshimura and Prudhomme (1988) have proposed a similar method for measuring V_s with two different shear gaps in which β is maintained constant. However, this restriction is not necessary for narrow gap viscometers that are analyzed in terms of τ_m and V/H. In this case, the slip velocity can be calculated from measured differences in the mean shear rates at constant shear stress with two different shear gaps:

$$V_s = \frac{(V/H)_1 - (V/H)_2}{1/H_1 - 1/H_2}, \ \tau_m = c \tag{6}$$

and the corrected shear rate is:

$$\left(\frac{V}{H}\right)^* = \frac{V}{H} - \frac{V_s}{H} \tag{7}$$

It should be noted that the shear stress and the Newtonian shear rate at R_b:

$$\tau_b = \frac{T}{2\pi R_b^2 L} \tag{8}$$

$$(\dot{\gamma}_b)_1 = \frac{2\Omega}{1 - \beta^2} \tag{9}$$

which are often used to interpret rheological data, cannot be used as consistency parameters. For example, in addition to being poor approximations of the Bingham Plastic, they show a trong dependency on the shear gap that could be misinterpreted as wall slip. In particular, the alculations in Fig. 2 show that the Newtonian shear rate is higher for the smaller gap (β = 96), which is the characteristic behavior of materials that exhibit wall slip.

.2 Couette viscometer experiments

he Couette viscometer used in these experiments was a Chan 35 (manufactured by Chandler ngineering), which is commonly used for drilling muds and oil-well cements. The outer cylin- er was the standard R1 rotor (R_c = 18.43 mm) that can be turned at 12 different speeds up to 00 RPM. For this viscometer, the torque is measured on the inner cylinder or stationary bob R_b), which was varied to provide different shear gaps. However, $\beta = R_b/R_c$ was kept close nough to 1.0 that the mean shear stress (τ_m), and mean shear rate (V/H) can be used as consis- ency parameters.

The Class H oil-well cements used in these experiments were prepared according to the American Petroleum Institute (API) Spec 10 (1982). This procedure involved initially mixing he water and cement (W/C = 0.38) for 15 seconds at 4000 RPM and then for 35 seconds at 2,000 RPM. After high-speed mixing, the cement slurry was stirred in an atmospheric con- istometer for 20 minutes at 150 RPM. Immediately after removing the slurry from the consis- ometer, rheological measurements were made with the concentric cylinder viscometer at ecreasing shear rates using the smaller shear gap (H = 0.70 mm). The method used for naking these measurements involved conditioning the sample at the highest shear rate for 60 econds, and then making measurements at progressively lower shear rates after only 20 sec- nds. The viscometer was then emptied and a fresh sample of cement, which was being stirred n the consistometer, was measured with the larger gap (H = 1.78 mm). The total time elapsed rom the initial addition of water to the end of the test was approximately 35 minutes. It is mportant to mention that the torque measured by the atmospheric consistometer did not show a neasurable increase over a period of one hour at room temperature (22-24°C). Consequently, it an be concluded that the viscosity of this cement does not increase significantly over the test eriod required to measure both shear gaps.

Rheological measurements of a bentonite clay/water slurry (10 wt% Aquagel) are shown in Fig. 3. Although the flow curve for this slurry is characteristic of a material with a yield value, here is no significant separation in data obtained with different shear gaps. Consequently, it an be concluded that this particular slurry does not exhibit a measurable degree of wall slip nder the conditions of this experiment. On the other hand, the cement slurry shown in Fig. 4 xhibits a very large degree of wall slip as evidenced by the large separation in the mean shear ates at constant shear stress. In particular, at τ_m = 20 Pa, the correction for wall slip (V_s/H), vhich was calculated from differences in the measured shear rates, was as large as the mea- ured shear rate (V/H). This critical condition of 100% slip is indicative of a yield value of approximately 20 Pa. While none of the data for $\tau_m < 20$ Pa contain any useful rheological nformation, we would not be aware of this fact if we had not made measurements with two dif- erent gaps. For example, the uncorrected data in Fig. 4 is indicative of a material that approaches a limiting Newtonian viscosity at low shear rates instead of a material with a yield value.

The relative importance of slip to shear flow is given by the ratio of the slip velocity to the otal velocity across the shear gap (V_s/V). This quantity is referred to as % slip, and is shown n Fig. 5 for several different cement slurries (Class H, W/C = 0.38). The data that were reported earlier (Mannheimer, 1983) indicate that 100% slip was reached for $\tau_m \leq 10$ Pa;

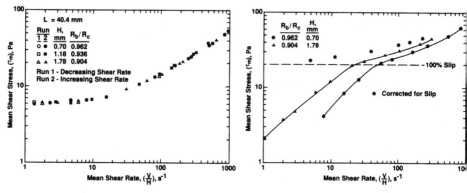

Fig. 3. Couette data for clay slurry

Fig. 4. Couette data for cement slurry

however, for $\tau_m > 24$ Pa no measurable slip could be detected with four different shear gaps ($0.71 \leq H \leq 2.42$ mm). The results for the four more recent slurries in Fig. 5, which were made from a different batch of Class H cement, include data for the slurry shown in Fig. 4. For these four slurries, the condition of 100% slip was reached at 15-20 Pa, and while % slip decreased for $\tau_m > 20$ Pa, it was still a significant factor at $\tau_m = 50$ Pa. Measurements of the free water content of these slurries (as determined by API Spec 10) produced values in the range of 5.6-6. mL, which exceeded the maximum allowable value of 3.5 mL. Since it was suspected that the high free water content was responsible for the large amount of slip, the free water was reduced to 3 mL by the addition of 0.3 wt% bentonite (Aquagel), but the measurements in Fig. 5 show that instead of reducing wall slip, the addition of 0.3% bentonite only increased the yield value of the cement slurry.

2.3 Twin-tube viscometer experiments
The twin-tube viscometer, shown schematically in Fig. 6, makes it possible to obtain rheological measurements with two different size tubes ($R_1 < R_2$) simultaneously (Gleible and Windhab,

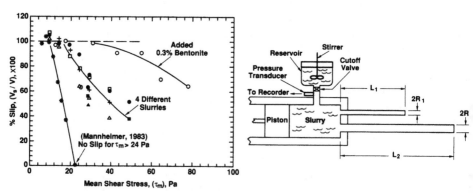

Fig. 5. Effect of shear stress on slip

Fig. 6. Schematic of twin-tube viscometer

152

1985). Consequently, samples of cement with the same shear history can be measured with two different shear gaps in a relatively short time. The length of the tubes were chosen so that $(L/R)_1 = (L/R)_2$, and since both tubes are connected to a common reservoir, the pressure drop (ΔP) will be the same across both tubes. Consequently, if kinetic energy and entrance effects can be neglected, $(\tau_w)_1 = (\tau_w)_2$, and any difference in the Newtonian shear rates ($4V/R$) for the two different tubes can be attributed to wall slip (Fredrickson, 1964). Typically, the flow rate was measured for one tube by collecting and weighing a sample for a specified time. The flow rate for the second tube could then be obtained by subtracting this value from the total flow rate at a specified pump setting.

The results of duplicate twin-tube experiments for a non-Newtonian clay slurry (8 wt% Aquagel) are presented in Fig. 7. Except for a few higher than normal points for the smaller tube ($R = 1.27$ mm) in run 2, these data show no dependency on tube size. Consequently, it can be concluded that this slurry does not exhibit a measurable degree of wall slip under the conditions of this experiment. Furthermore, the best fit of these data to the Buckingham-Reiner equation for the flow of a Bingham Plastic corresponded to $\tau_o = 7.7$ Pa and $\mu_{pl} = 16.3$ mPa s (see solid curve in Fig. 7). However, measurements of a Class H cement slurry ($W/C = 0.38$) prepared according to API Spec 10 showed a very strong dependence on the size of the tube that is indicative of wall slip (see Fig. 8).

It is important to note that the larger tube in Fig. 8 is not the one that was used for the clay slurry in Fig. 7. The reason for this change was that anomalous corrections for slip ($4V_s/R \gg 4V/R$) were obtained when the two tubes used for the clay slurry in Fig. 7 were used with cement slurries. The reason that these slip corrections were too large was discovered when a small section of each tube was cut in half lengthwise and surface roughnesses measured with a profilometer. These measurements indicated a surface roughness of 1.7 μm for the smaller tube ($R_1 = 1.27$ mm) and 3.4 μm for the larger tube ($R_2 = 1.95$ mm). Since the higher surface roughness would tend to reduce the effects of wall slip, the Newtonian shear rates would be shifted (at constant shear stress) to lower values for the larger tube. Therefore, the slip velocities and the corrections for slip would be too large. Just the opposite would be expected if the smaller tube were a lot rougher than the larger tube (i.e., the slip corrections would probably be too small). Consequently, new tubes were chosen that were significantly different in size (1.27 mm and 2.4 mm), but which had similar surface roughnesses (1.7 μm and 2.2 μm). Except for the

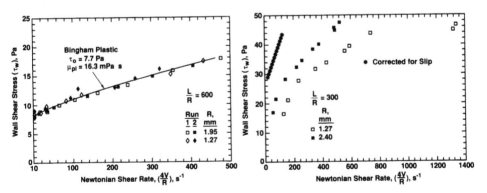

Fig. 7. Twin-tube data for clay slurry   Fig. 8. Twin-tube data for cement slurry

anomalous data for the smaller tube at the two highest values of τ_w in Fig. 8, the slip velocities, calculated from differences in Newtonian shear rates at constant stress, were used to correct the data for wall slip for $28.6 \leq \tau_m \leq 42$ Pa, but for $\tau_m < 28$ Pa, the flow was essentially 100% slip.

Since it is possible for the rheological properties of a cement slurry to change with shear history and time, Couette viscometer measurements were made of the same cement slurry before and after the twin-tube experiment shown in Fig. 8. The results of these three runs, all of which were corrected for wall slip, are compared in Fig. 9. The fact that run 3 is slightly higher than run 1 indicates the apparent viscosity ($\mu_a = \tau/\dot{\gamma}$) of this cement slurry increased slightly (10-15%) with the age of the slurry, which was approximately 35, 80, and 100 minutes after runs 1, 2, and 3, respectively. The twin-tube data in Fig. 9 tend to fall between runs 1 and 3 with the Couette viscometer at the higher shear stresses, but below run 1 at the lower shear stresses. This discrepancy can be explained in part by the corrections for shear dependency:

$$(\dot{\gamma}_w)^* = \left(\frac{3n'+1}{4n'}\right)\left(\frac{4V}{R}\right)^*$$

(10)

$$n' = \frac{d \ln \tau_w}{d \ln (4V/R)^*}$$

(11)

which were very large for $\tau_w < 35$ Pa. Thus, even small errors in estimating n' can result in a large error in the shear rate. Nevertheless, the important conclusion to be drawn from the results in Fig. 9 is that rheological measurements of a cement slurry that exhibits wall slip can be made with different viscometers, and still be in reasonably good agreement if the data are corrected for slip.

The slip velocities, which were used to calculate wall slip corrections for runs 1-3 in Fig. 9, are presented in Fig. 10. The slip velocities calculated from the twin-tube experiment (run 2) show almost the same dependency on shear stress as run 1 with the Couette viscometer. Furthermore, the slip velocities for shear stresses above 20 Pa were much lower in run 3 with the Couette viscometer. These results suggest that after approximately 80 minutes there is a

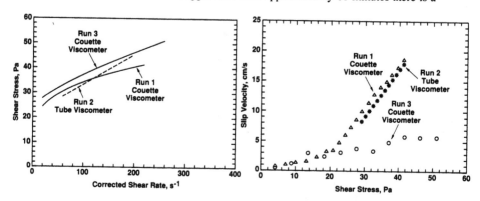

Fig. 9. Couette and tube data for a cement slurry   Fig. 10. Slip velocities for a cement slurry

dramatic decrease in wall slip, which might be related to the adsorption of free water, and which could be checked by measuring the free water of aged samples.

3 Large-scale experiments

In order to conduct experiments on laminar-turbulent transition of cement slurries at relatively low shear rates, a large-scale pipe-flow facility was constructed (see Fig. 11). This facility consisted of a stirred reservoir (1500 L), a variable speed pump (0-16.7 L/s, Moyno), a mass flowmeter (0-53 kg/s, Micromotion), a pulsation dampener, and a test section consisting of 14.3 m of stainless steel pipe (D = 50.8 mm). The pressure gradient was measured with a differential pressure transducer, which was connected to pipe taps 4.9 m apart and approximately 162 pipe diameters downstream of any significant flow disturbance. The lines connecting the pressure transducer to the pipe taps were filled with water and could be back-flushed if they became plugged with cement.

Benchmark experiments with Newtonian liquids showed: 1) viscosities calculated from laminar flow data were in good agreement with laboratory viscometers, 2) laminar-turbulent transition occurred at the expected values of the Reynolds number (2000-2100), and 3) the friction factors ($f = 2\tau_w/\rho V^2$) for fully turbulent flow were representative of smooth pipe (Park, et al., 1989).

Because the large-scale experiments were conducted at higher temperatures (38-39°C) than the laboratory experiments (22-23°C), it was necessary to use a retarding agent (HR-4, Halliburton) at a concentration of 0.2 wt% based on the weight of the cement. The results for a Class H cement (W/C = 0.38) are shown in Fig. 12. Measurements made at decreasing and increasing shear rates showed no hysteresis. Therefore, it can be concluded that no significant changes in rheological behavior of the slurry had taken place due to shear or turbulence. Furthermore, no plugging of the pipe taps by cement had occurred during the test. The laminar data in Fig. 12 were fitted to the Buckingham-Reiner equation for laminar pipe-flow of a Bingham Plastic by a nonlinear regression technique (see solid curve in Fig. 12) to obtain τ_o = 16.1 Pa and μ_{pl} = 55.5 mPa s. The rheological properties of several slurries (both clay and cement), which were determined by fitting laminar pipe-flow data to the Bingham Plastic model, are summarized in Table 1.

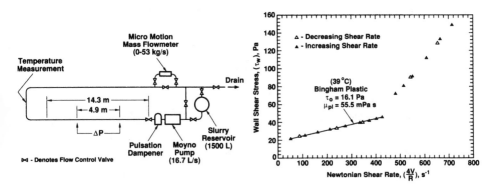

Fig. 11. Large-scale pipe-flow facility   Fig. 12. Large-scale pipe flow of cement slurry

Table 1. Rheological properties determined from laminar pipe-flow data

Composition	ρ (kg/L)	τ_o (Pa)	μ_{pl} mPa s	N_{He}
8 wt% Clay	1.05	6.5	27.4	2.6 x 10⁴
9 wt% Clay	1.07	15.9	38.8	3.1 x 10⁴
10 wt% Clay	1.08	34.9	70.1	2.1 x 10⁴
11 wt% Clay	1.09	55.4	82.4	2.4 x 10⁴
W/C = 0.37	2.03	16.1	60.9	2.4 x 10⁴
W/C = 0.38	2.03	16.1	55.5	2.9 x 10⁴

Despite the wide range of rheological properties, all of the slurries in Table 1 were characterized by approximately the same Hedstrom number ($N_{He} = \tau_o \rho D^2 / \mu_{pl}^2$). Consequently, the friction factor for laminar pipe flow should be a unique function of the Reynolds number based on the plastic viscosity ($N'_{Re} = DV\rho/\mu_{pl}$) as predicted by Hedstrom (1952). This prediction is confirmed by the pipe-flow data in Fig. 13 for $150 \le N'_{Re} \le 5000$, which represents measurements for all of the slurries in Table 1. Furthermore, the critical Reynolds number for laminar-turbulent transition (N_{Re})$_c$ should be in the range of 4600 to 5000 for ($2.1 \times 10^4 < N_{He} < 3.1 \times 10^4$) as predicted by Hanks (1963), which is also confirmed by the results in Fig. 13. Finally, the friction factors for fully turbulent flow of all these slurries fall only slightly below the values predicted by the Blasius equation for Newtonian liquids in smooth pipe. Based on these results, it is evident that we can predict friction factors for laminar flow, laminar-turbulent transition and turbulent friction factors of clay and cement slurries from rheological properties measured with large-diameter pipe. Unfortunately, predictions based on rheological properties made with a laboratory viscometer were not nearly as useful.

In particular, approximately one liter of cement slurry ($W/C = 0.37$) was taken during the large-scale experiments and immediately measured with a Couette viscometer using two different shear gaps to check for wall slip. These measurements were obtained at decreasing shear rates according to API Spec 10. The results in Fig. 14 show that the narrow gap data

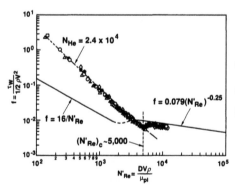

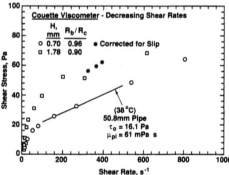

Fig. 13. Friction factor data for clay and cement slurries

Figure 14. Couette and large-scale pipe data for cement slurry

156

$H = 0.70$ mm) are in good agreement with pipe data (see solid line in Fig. 14) for shear rates in the range of 100 s^{-1} to 500 s^{-1}. The large separation in the mean shear rates at constant shear stress would normally be explained in terms of wall slip; however, the data for the wide gap $H = 1.78$ mm) exhibited an anomalous hump near 200 s^{-1} that was not evident with cement slurries blended and tested in the laboratory. Furthermore, while a few corrections for slip could be made for the data above this hump, the corrections for slip only made matters worse.

It was suspected that these anomolous results might be caused by a cement plug that was not at shear equilibrium. However, similar results were obtained when a second sample was withdrawn after increasing the water content to $W/C = 0.38$. In particular, two runs at decreasing shear rates and one run at increasing shear rates gave almost the same results (i.e., close agreement between the large-scale data and the Couette data obtained with $H = 0.7$ mm, and a wide separation in gap size that would suggest wall slip. Furthermore, data that were corrected for wall slip overestimated the apparent viscosity by approximately 30% at 500 s^{-1} and by 100% at 100 s^{-1}. The reasons for these poor results have not been resolved, but may be related to the retarder that had not been used in laboratory experiments.

Conclusions

- Rheological measurements of cement slurries made with different laboratory viscometers can give comparable results if the data are corrected for wall slip. However, special attention must be taken with tube viscometers to ensure that different size tubes have approximately the same surface roughness.

  Theoretical predictions of the critical Reynolds number for laminar-turbulent transition of clay and cement slurries were in good agreement with measured values provided these predictions were based on rheological properties measured from laminar flow data in a large-scale pipe viscometer.

- A comparison of Couette viscometer measurements of cement slurries taken from the large-scale experiments showed good agreement with data obtained with the smaller gap, but not with the larger gap. Although this difference can be used to correct the data for slip, the anomalous appearance of the data obtained with the larger gap suggests that some factor other than slip is involved. Consequently, further research is required to resolve these questions.

References

API specification for materials testing and for well cements. (1982) **SPEC 10,** first edition.
Bannister, C.E. (1980) Rheological evaluation of cement slurries—methods and models. **SPE 9284.**
Bhatty, J.I. and Banfill, P.F.G. (1982) Sedimentation behaviour in cement pastes subjected to continuous shear in rotational viscometers. **Cement and Concrete Research,** 12, 69-78.
Cheng, D.C. (1984) Further observations on the rheological behaviour of dense suspensions. **Powder Technology,** 37, 255-273.
Frederickson, A.G. (1964) **Principles and Applications of Rheology,** Prentice-Hall, Englewood Cliffs, N.J., pp. 19 & 194.
Gleible, W. and Windhab, E. (1985) The 'twin capillary', a simple device to separate shear- and-slip-flow of fluids. **Expts. in Fluids,** 3, 177-180.

Hanks, R.W. (1963) The laminar-turbulent transition for fluids with a yield stress. **AIChE J.**, 9, 306.

Hedstrom, B.O.A. (1952) Flow of plastic materials in pipes. **Ind. & Engr. Chem.**, 44, 651-656.

Mannheimer, R.J. (1982) Rheological evaluation of cement slurries. **SwRI Report No. 6836.**

Mannheimer, R.J. (1983) Effect of slip on flow properties of cement slurries. **Oil and Gas J.**, December, 144-147.

Mannheimer, R.J. (1985) Flow characteristics of slurries at shear stresses near the yield value. **Proceedings, 10th International Technical Conference on Slurry Technology,** (ed B.A. Sakkestad), Slurry Technology Association, Washington, D.C., pp. 123-133.

Mannheimer, R.J. (1988) Laminar and turbulent flow of cement slurries in large diameter pipes. **SwRI Report No. 8983.**

Mooney, M. (1931) Explicit formulas for slip and fluidity. **J. Rheology**, 2, 210.

Park, J.T. Mannheimer, R.J. Grimley, T.A. and Morrow, T.B. (1989) Pipe flow measurements of a transparent non-Newtonian slurry. **J. Fluids Engrg**, 111, 331-336.

Yoshimura, A. and Prudhomme, R.K. (1988) Wall slip corrections for Couette and parallel disk viscometers. **J. Rheology**, 32, 53-67.

6 Acknowledgements

This research was partially funded by Southwest Research Institute and the American Petroleum Institute. The author wishes to thank Mr. P. Gutierrez and Mr. R. Railsback for conducting the rheological measurements with laboratory viscometers and Mr. T. Grimley for performing the large-scale experiments. Special thanks are also offered to Ms. A. Tate for typing this paper and Mr. J. Pryor and Ms. B. Ford for helping to prepare the final manuscript.

17 HIGH TEMPERATURE AND HIGH PRESSURE RHEOLOGY OF OILWELL CEMENT SLURRIES

D.S. KELLINGRAY, C. GREAVES and R.P. DALLIMER
BP Research, Sunbury Research Centre, UK

Summary
Current test procedures limit the maximum temperature at which the
rheological properties of oil well cements can be determined to
85°C. When deep hot wells are drilled these measurements may give a
poor prediction of the cement slurry properties in the well bore,
increasing the chance of a cementing failure.

Pressure drops calculated from these rheological properties may be
different to those encountered during displacement. Inaccurate
predictions of pressure drops may result in the choice of an
inappropriate cement displacement rate, which compromises the
primary cementing operation. Accurate measurement of the rheology
of oil well cement slurries will be important to computer models of
cement placement, many of which are being developed to optimise
cementing operations. The Bingham yield point has been used as an
indicator of cement settlement behaviour, errors in its measurement
will result in pumping cement slurries which may settle in the
wellbore.

Pressurised consistometers are used to determine the pumping time
of oil well cement slurries at high temperature (200°C) and pressure
(96.5 MPa). A top drive consistometer has been modified to obtain
information on the rheological properties of cement slurries. The
shear rate shear stress data has been fitted to a Bingham plastic
model, enabling the plastic viscosity and yield point to be
determined at high temperature and pressure.

The investigation concludes that the modified consistometer can be
used to investigate oil well cement rheology under simulated
wellbore conditions. The suitability of the API Specification 10
procedure will be formulation dependent. A generalised model for
high temperature oil well cement rheology is not yet possible
because of the wide range of additives used in oil well cement
slurries.

1 Introduction

1.1 Background
When deeper, hotter wells are drilled, oil well cement slurries are
required which have the following properties:

-controlled water loss to the formation,
-no settlement at high temperatures,
-low frictional pressure drops during pumping,
-displacement of the drilling fluid in the hole.

The design of oil well cement slurries to meet these criteria has
resulted in highly complex cement slurry formulations. A slurry
design suitable for cementing a reservoir with a geothermal
temperature of 200°C would contain:

-oil well cement
-silica flour
-cement retarder (e.g. calcium lignosulphonate)
-cement retarder aid (e.g. borax)
-cement dispersant (e.g. naphthalene sulphonate)
-cement fluid loss aid (e.g. cellulosic)
-antifoam
-water.

The temperature that the slurry would achieve during displacement
would be somewhat lower than the geothermal temperature but would
still be significantly above 85°C. This lower temperature is known
as the Bottom Hole Circulating Temperature (BHCT).
Pressure is known to have an effect on the rheology of water based
drilling fluids(2). As increased pressure accelerates cement
hydration (3), it may also be expected to have an effect on cement
rheology.

1.2 Determining cement slurry rheology

The technique adopted throughout the oil industry is described in
API Specification 10 (1). This involves mixing the cement slurry to
a specified procedure in a Waring blendor. The slurry is then
conditioned for 20 minutes (in an atmospheric consistometer(1)) at
the predicted downhole temperature or 85°C, whichever is the lowest.
The sample is then placed in a rotational viscometer capable of
measuring shear stress at shear rates between 1.5 and 511s^{-1}, the
viscometer should have the following dimensions:

Rotor sleeve   Bob
Inside diameter = 36.83mm Diameter   = 34.49mm
Total length   = 87.0 mm Cylinder length = 38.0 mm

The viscometer should be direct indicating with the outer cylinder
being driven at a constant velocity, the torque being measured on
the inner cylinder/bob(commonly a Fann 35 is used). Since the
equipment can only be used at atmospheric pressure the maximum test
temperature is 85°C. This is a poor simulation of the conditions
the cement slurry may experience in deep hot wellbore.
To determine the pumping time of oil well cement slurries for high
temperature and pressure (HT/HP.) conditions, pressurised
consistometers are used (1). Modifying a consistometer to enable
torque and rotational speed to be recorded would be a simple way of

investigating the HT/HP rheology of oil well cement. The Nowsco
system was chosen because the rheology package is available as a
retrofit to existing equipment. It consists of a pressure vessel, a
magnetic drive and a bob matched to the pressure vessel(4). The gap
between rotor and stator is 3.75mm and the bob height is 94.8mm.
The motor is capable of rotating the bob at 0-900rpm (maximum shear
rate $2500s^{-1}$), the maximum shear stress is 1200 Pa.
   The pressure vessel sits in a heating jacket and is pressurised
hydraulically using an inert transformer oil. The test procedure is
to insert the bob into the pressure vessel, then to add sufficient
cement to just cover the top of the bob. The surface of the cement
is then flooded with oil.

1.3 Objectives of the investigation
This investigation had four objectives:

   -to develop a technique for measuring the HT/HP rheology of oil
   well cements.
   -to determine the level of accuracy to which HT/HP rheology could
   be determined.
   -to determine what differences there are between API Specification
   10 rheology and that determined under simulated wellbore
   conditions.
   -to define how important the determination of the rheological
   properties at HP/HT conditions was to predicting the frictional
   pressure drop during cement displacement.

2 Development and Calibration

2.1 Equipment development

2.1.1 Bearings
The equipment was originally supplied with a bob with PTFE tip on
which it rotated. In the first series of tests unexpected increases
in viscosity were observed with increasing pressure. This was
identified as wear and deformation of the PTFE tip. Replacement
with a steel tip reduced the problem.

2.1.2 Bob
The bob originally supplied was solid stainless steel, giving a
narrow gap at the bottom of the rheometer which was found to jam
with solids if the slurry showed any signs of settlement. The
weight of the bob was also in part responsible for the problem with
the PTFE tip. The bob supplied was replaced with a hollow bob which
had an increased clearance between the base of the pressure vessel
and the bob. Therefore subsequent measurements were made with the
hollow stainless steel bob.
   The hollow bob is more difficult to model theoretically as the end
effects and the effects around the shaft and spokes connecting to
the cylinder are highly complex. However as high levels of

161

precision were inappropriate during this investigation, these second order effects were ignored.

2.2 Calibration

2.2.1 Water
To determine the friction due to mechanical bearings on the shaft, the viscosity of water was measured at various pressures. The observed increase in viscosity can only be attributed to the friction at the bearings on the shaft. The effect is summarised in Table 1.

2.2.2 Calibration oil
An oil which had previously been characterised in two different HT/HP. rheometers (Huxley Bertram and a capillary type viscometer, neither of which were suitable for cement slurries) was tested in the modified consistometer. The results reported in Table 2 (Figure 1) indicate a good correlation with the data from the other rheometers.

Table 1. Effect of Pressure on Friction on Bearing for Shear Rates Below 511 S^{-1} for HT/HP. Rheometer Containing Water

Shear Rate s^{-1}	Pressure (MPa) 0	20.3	40.5	60.8	81.1
	indicated shear stress (Pa)				
511	0	1.5	2	3.5	5.5
341	0	1	1.5	2	3
171	0	0	0	0.5	1
102	0	0	0	0	0.5
51	0	0	0	0	0

Table 2. The Effect of Pressure on the Viscosity of a Silicone Calibration Oil at 25°C

Pressure (MPa)	Average viscosity determined in 2 different HP (Pa s)	Viscosity determined in the modified consistometer rheometers (Pa s)
0	0.017	0.020
20.3	0.027	0.028
40.5	0.036	0.038
60.8	0.043	0.044
81.1	0.075	0.069

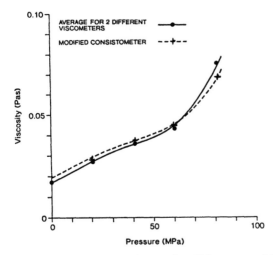

Fig.1   Comparison of viscosity of silicone calibration oil using
two different high pressure rheometers and the modified
consistometer.

3 EXPERIMENTAL

3.1 Oil Well Cement Rheology
Two oil well cementing formulations were investigated:

(a) a slurry design suitable for cementing intermediate casing
containing cement, silica flour and retarder mixed with water to a
density of 1920 Kg m^{-3}.
(b) a slurry suitable for cementing across a reservoir, containing
cement, silica flour, fluid loss aid, dispersant and retarder
mixed in distilled water at 1920 Kg m^{-3}.

Both slurries were tested using the following procedure.

-mix the slurry according to API Specification 10
-condition the slurry at 85°C in an atmospheric consistometer.
-determine the rheological properties at 85°C using the API and
the HT/HP. rheometers (unpressurised), determining shear stress at
shear rates of, 511, 341, 171, 102, 51 s^{-1}. The value's were
steady state values achieved after 20 seconds at the defined shear
rate (after an initial 60 seconds at 511s^{-1}). The plastic
viscosity and yield point were determined from a best fit of the
data to a Bingham plastic model.
-using HT/HP. rheometer, pressurise sufficiently to prevent water
boiling from the slurry and increase temperature to BHCT,
monitoring apparent viscosity.
-measure rheological properties at BHCT.
-increase pressure in steps to simulated downhole pressure and
measure rheological properties.

a) RETARDED CEMENT SLURRY

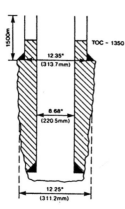

Top of cement @ 1350m

SPACER

Vol 60 bbls (9.54m³)
Plastic Viscosity = 0.02 Pas
Yield Stress   = 4.8 Pa
Density   = 1500 kg/m³

MUD

Plastic Viscosity = 0.04 Pas
Yield Stress   = 9.6 Pa
Density   = 1400 kg/m³

b) COMPLETION CEMENT SLURRY DESIGN

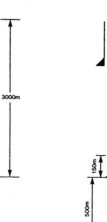

Fig.2   Oilwell   parameters   for   frictional   pressure   drop
determinations.

3.2 Determination of frictional pressure drop

The model wells and the properties of the drilling fluid and the
cement are summarised in Figure 2. The frictional pressure drops
were calculated using a computer program which uses models for
laminar (5) and turbulent (6) flow; the accuracy of the turbulent
predictions is considerably poorer than those in laminar flow. The
program uses an equivalent diameter approach to model turbulent
flow. This is defined as the o.d. of the inner pipe subtracted from
the i.d. of the outer pipe. The fluids are considered as Bingham
plastic fluids flowing through concentric pipes.

4 Results

4.1 Cement slurries

The plastic viscosity and yield stress determined using the API
Specification 10 procedure and the HT/HP rheometer are reported in
Table 3. The effect of temperature and pressure on the apparent
viscosity of the simple cement slurry containing cement/silica
flour/retarder and water is reported in Tables 4-5. The results for
the more complicated cement slurry formulation are reported in
Tables 6-8.

164

Table 3. Plastic Viscosity and Yield Point for a 1.92 SG Class G
Cement Slurry with Silica Flour and Retarder Determined
using a Pressurised and Unpressurised Rheometer

Rheometer	Pressure (MPa)	Temperature °C	Plastic Viscosity Pa.s	Yield Point Pa
API Spec 10	ambient	85	0.042	16
HP/HT	ambient	85	0.038	19
HP/HT	2.1	106	0.040	22
HP/HT	40.5	106	0.029	23
HP/HT	81.1	106	0.026	26

Table 4. Effect of Temperature on
Apparent Viscosity at
511 S^{-1} Measured AT 2 MPa

Temperature (°C)	Apparent Viscosity (Pa.s)
90	0.078
95	0.079
100	0.080
105	0.084
107	0.087

Table 5. Effect of Pressure on
Apparent Viscosity at
511 S^{-1} Measured at 106°C

Pressure (MPa)	Apparent Viscosity (Pa.s)
2	0.073
40.5	0.079
81.1	0.091

Table 6. Plastic Viscosity and Yield Point for a 1.92 SG Cement
Slurry Suitable for Completing an Oil Well

Rheometer	Pressure (MPa)	Temperature (°C)	Plastic Viscosity (Pa.s)	Yield Stress (Pa)
API Spec 10	ambient	85	0.353	11
HP/HT	ambient	85	0.291	9
HP/HT	1.0	85	0.299	9
HP/HT	1.0	119	0.162	4
HP/HT	40.5	119	0.162	4
HP/HT	81.1	119	0.164	4

Table 7. Effect of Temperature on Apparent Viscosity at 511 S^{-1} Measured at 1 MPa

Temperature ('C)	Apparent Viscosity (Pa.s)
86	0.331
92	0.307
97	0.282
106	0.244
111	0.212
116	0.188
119	0.174

Table 8. Effect of Pressure on Apparent Viscosity at 511 S^{-1} Measured at 119°C

Pressure (MPa)	Apparent Viscosity (Pa.s)
1	0.174
40.5	0.182
81.1	0.200

4.2 Frictional pressure drops

For the retarded slurry the equivalent circulating densities (ECD) (the sum of the frictional pressure drops during displacement and the hydrostatic pressure), were determined for a 3000m well described in Figure 2. The ECD of the completion slurry in a 3500m well described in Figure 2 were also determined. The ECDs have been determined for the bottom of the well and represent the maximum ECD during cement displacement.

5 Discussion

5.1 The effect of temperature and pressure on oil well cement slurry rheology

5.1.1 Rheological models

The choice of the Bingham Plastic model ignores the considerable debate about the most appropriate rheological model. In this paper it has been chosen for simplicity as the primary objective was to assess the importance of temperature and pressure on slurry rheology.

5.1.2 Effect of high pressure

The effect of pressure on the apparent viscosity at $511s^{-1}$ of the two slurries at BHCT is compared in Figure 3. There is a marginal increase with increasing pressure, this may be accounted for by the compression of the liquid phase and corresponding increase in solid fraction of the slurry.

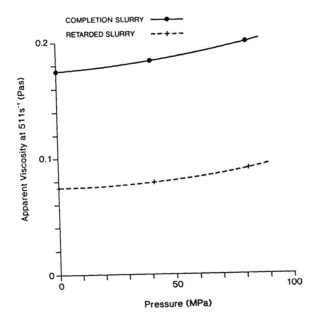

Fig. 3   Effect of pressure on the apparent viscosity of oil well cement slurries.

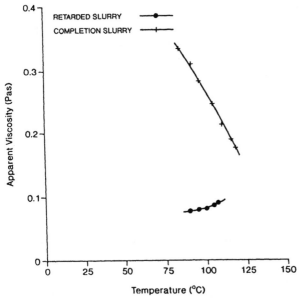

Fig. 4   Effect of temperature on rheology of oil well cement slurries.

5.1.3 Effect of high temperature

There are major differences between the response of the two cement slurry designs to increased temperature; these are compared in Figure 4. The simple slurry containing retarder has little dependence on temperature. Therefore it can be concluded that rheology of simple slurries can be determined using the current API Specification 10 Operating rheology test. However the slurry which would be used to cement the reservoir section exhibits significant thermal thinning. The plastic viscosity reduces to 50% of the value determined using the API Specification 10 procedure. The yield stress determined at the simulated downhole conditions is 33% of the value from the API Specification 10 test. This reduction in yield stress may have a significant impact on the settling of the cement slurry in the Wellbore.

5.2 Effect of determining cement rheology at HT/HP on frictional pressure drop

5.2.1 Retarded slurry

There is only a small difference in the predicted ECD resulting from using HT/HP rheological properties. A small increase is indicated however this is unlikely to make a significant difference to the achievable displacement rate or success of the job.

5.2.2 Completion slurry

As a consequence of the viscosifying action of the fluid loss aid, the slurry has a very high ECD. Testing the slurry at HT/HP conditions makes a significant difference to the ECD, at 0.636 m3/min (4 bbl/min) it is reduced by 7.6%. Additionally the flow rate at which turbulence occurs is reduced by 50%, from 6.677 m3/min to 3.17 m3/min, although these flow rates are not practical for oil well cementing operations, the ability to pump faster will improve the quality of the cement job. If ECD was limited to 2.2SG, the HT/HP rheology measurements suggest a maximum flow rate of 1.27-1.43 m3/min (8-9 bbl/min), the API Spec 10 rheology indicates a maximum flow rate of 0.63-.79 m3/min (4-5 bbl/min).

6 Conclusions

The following conclusions have been drawn in response to the objectives defined in Section 1.3:

-with some simple modifications the Nowsco rheology package can be used to assess the high temperature rheological properties of oil well cement slurries.
-under ambient conditions there was good agreement between the Nowsco rheometer and a rheometer meeting API Specification 10. Except at very high pressure (81.1 MPa) there was good agreement between the Nowsco and two different high pressure rheometers.
-the differences between the rheology determined at 85°C and that at the anticipated wellbore temperature was found to be

Table 9. Effect of Determining Rheology at HT/HP. Conditions on the Effective Circulating Density at Bottom of Well Effective circulating density (SG equivalent)

Cement displacement rate (bbl/min)	Retarded slurry API Spec	HT/HP.	Completion slurry Spec	HT/HP.
0.2	1.83	1.86	1.82	1.77
0.3	1.84	1.87	1.91	1.82
0.6	1.85	1.88	2.08	1.93
1.3	1.87	1.9	2.43	2.14
1.6	1.88	1.91	2.6	2.24
2.4	1.90	1.92	3.02	2.50
3.2	1.93	1.94	3.44	2.77

formulation dependent. For a simple retarder cement slurry there was not a significant change with increasing pressure. However the slurry containing a fluid loss aid showed significant thermal thinning, the plastic viscosity at 119°C was 50% of the value at 85°C, the yield point was 33% of the value at the lower temperature.
-for complex slurries containing polymers which reduce in viscosity with increasing temperature, measuring the rheological properties at simulated wellbore conditions will have a major effect on the frictional pressure drops predicted during displacement and the choice of cement displacement rate. Simple slurries which do not exhibit a significant thinning with increasing temperature can be tested using the API Specification 10 procedure without significantly compromising the cement displacement.

7 References

"Specification for Materials and Testing for oil well cements". API Specification 10 Third Edition, August 87.
"High-Temperature, High-Pressure Rheology of Water-Based Muds". N.J.Alderman et al. SPE 18035 1988
"The effects of pressure on the set properties of cements with various additives". T.D Drescher and A.S.Metcalf. 25th Annu. Southwestern Petrol. Short Course Proc. pp13-19,1978.
PC-10 Rheology Accesory Instruction Manual. Nowsco Well Service Ltd.
"Slurry and Suspension Transport". W.M.Laird. Ind. Eng. Chem. vol. 49, no. 1 Jan. 57 pp 138-141.
"A new Analysis of Turbulent Flow of Non Newtonian Fluids". A.D.Thomas and K.C.Wilson. Can. Jour. Chem. Eng. vol. 63, Aug. 85 pp 539-546.

18 VISCOELASTIC PROPERTIES OF OILFIELD CEMENT SLURRIES

A. SAASEN, C. MARKEN and N. BLOMBERG
Rogaland Research Institute, Stavanger, Norway
J. DAWSON and M. ROGERS
BJ-Titan, Tomball, Texas

Abstract
Viscoelastic properties have been demonstrated in cement slurries
which have their application in oil wells. These properties were
found for both neat and polymer containing slurries. Over the
limited range evaluated, both the elastic and viscous parameter
values increased with the combination of increased pressure and
temperature.
Keywords: Viscoelasticity, Oscillating Rheometer, Cement Slurries,
Oil Wells.

1 Introduction

A successful completion of an oil/gas well depends on good primary
cementing of the casing. Primary cementing is done to establish seals
between the casing and the formation in order to prevent fluid or gas
migration between different zones and/or the surface. The cement
shall also support the weight of the casing and ensure correct casing
position.

Successful primary cementing requires control of the cement
rheology. Traditionally Bingham and power-law models have been used.
Presently the use of power-law models that are based on anticipated
shear rate ranges is getting more common. Normally, when using the
Bingham model, the yield point is kept as low as possible. Similarly,
the power-law index, n, is kept between 0.9 and 1.0. Other
rheological aspects of slurries, except for tixotropic behaviour,
have received little attention.

The present work demonstrates viscoelastic behaviour of cement
slurries which have their application in the oil industry.
Viscoelastic properties may be important for the cement displacement
efficiency, especially while cementing in narrow and irregular
annular hole sections of an oil well. Previous work, Chow et al.
(1988) and Cooke et al. (1988), has related viscoelastic parameters
to gel properties for both flowing and curing slurries. Chow et al.
(1988), investigated the use of viscoelastic parameters as an
alternative to consistometer studies. In that study the viscoelastic
behaviour of class H cement slurries were studied in connection with
investigations on gel properties and hydration times. The present
study demonstrate that the cement slurries exhibit viscoelastic

behaviour also in the non-gelled state. The significance of
viscoelasticity for the cement slurry flow remains beyond the scope
of this work.

2 Viscoelastic parameters

One of the most common ways to quantify viscoelastic properties of
fluids and mixtures is through the use of an oscillating rheometer.
It is common practice to measure viscous parameters by the use of a
concentric cylinder rheometer where the outer cylinder is rotated at
a specified shear rate. The torque measurement on the inner
cylinder, resulting from this shear, is then transformed to shear
stress and viscous values. If an oscillating outer cylinder is used,
the measurement is performed differently, see for example Chow et al.
(1988) and Bird et al. (1987). With this technique the outer
cylinder oscillation has the form $\gamma^0\cos(\omega t)$. The inner cylinder
measures an oscillating stress which differs from the outer cylinder
oscillation through an amplitude and phase change. This difference is
used to calculate the linear viscoelastic properties G' and G''. The
shear stress, τ, is given by equation 1;

$$\tau = G' \gamma^0 \sin(\omega t) + G'' \gamma^0 \cos(\omega t) \tag{1}$$

where ω is the angular velocity, t is the time and γ^0 is the strain
amplitude. G' accounts for the stored elastic energy of the fluid,
while G'' accounts for the dissipated energy as shown by Bird et al.
(1987). For a Newtonian fluid (or slurry) $G''/\omega=\mu$, the viscosity,
and $G'=0$. For an elastic fluid an increase in elastic properties is
represented by a increased G'.

Typically, at a small and constant amplitude, a frequency sweep
will show G' increasing with frequency. A strain sweep, for small
strains, will also give valuable information. A gelled fluid, or a
fluid with a yield point, should exhibit no decrease in G' with
strain at small strain values. A decrease indicates a immediate
breakdown of the static fluid structure. Typically for a non-gelled
fluid, the fluid structure is always changing with strain and G'
should decrease with strain for low strain values. When the strain
becomes large it is difficult to get valuable information about
viscoelastic properties due to non-linear effects.

3 Experimental conditions

Each cement slurry used in this study was made from Class G cement
weighted to 1.89 kg/m^3. All solid additives, except those in Cement
A, were blended with the cement powder before addition to the water.
Liquid additives were added to the water just prior to the addition
of cement solids. Each slurry was formulated with the properties
necessary to do an oil well cement treatment. As a result, the co-
additive package differed in each slurry making direct comparison of
the data difficult. The four slurries were prepared and preconditioned in accordance to the American Petroleum Institute (1986).

Table 1. Fann 35 readout for the shear stress of the cement slurries

Shear Rate (rpm)	Cement A (dial reading)	Cement B (dial reading)	Cement C (dial reading)	Cement D (dial reading)
600	245	227	65	239
300	136	137	27	127
200	94	104	18	86
100	49	67	10	46
6	4	40	3	3
3	2	40	3	2

Table 2. Viscosity and shear rate for the cement slurries

	Cement A	Cement B	Cement C	Cement D
Shear rate (s^{-1})	Visc. (mPa·s)	Visc. (mPa·s)	Visc. (mPa·s)	Visc. (mPa·s)
1022	123	114*	33	120
511	136	137	27	127
340	141	156	27	129
170	147	201	30	138
10.2*	200	2000	150	150
5.1*	200	4000	300	200

* At these shear rates the viscosity values contain errors in the order of 100 mPa·s.

The cement solids were added within 15 sec. to the water sheared at 4000 rpm on a Waring blender. After the addition was completed, the shear rate was increased to 12,000 rpm for 35 sec. The slurry was then preconditioned in a Chandler Atmospheric Consistometer for 20 minutes at test temperature.

Following the preconditioning, the slurry viscosity was measured on a Fann 35 rheometer at ambient temperature (see Table 1). Stresses were recorded using 1022, 511, 340, 170, 10.2 and 5.1 s^{-1} (see Table 2). Afterwards, the slurry was syringed into the 18 mm radius cup of a Rheometrics Pressure Rheometer. The other rheometer tool used was a 17 mm radius bob resulting in a 1 mm shear gap. The samples were then pressured to 0.69 MPa for ambient temperature tests and 2.76 MPa for the 67^0C tests. The samples were then subjected to frequency and strain sweep measurements. All of the frequency sweep measurements, except for the Cement C tests at 0.69 MPa, were done at a strain of 30% of the annular gap. This latter test was performed at an amplitude of 50% of the annular gap.

As is the practice, the following weight percent of the components are based on the dry weight of cement solids. The cement compositions are shown below:

Cement A   44.32% fresh water, 1.0% polyvinyl acetate (PVA)
polymer, 0.3% naphtalene sulfonate resin as a
dispersant, 0.2% sodium metasilicate and 0.3% of a
lignosulfonate as a retarder. The polyvinyl acetate
polymer was hydrated in the water for about 12 hrs.
prior to addition of cement solids.

Cement B   46.75% seawater, 0.75% polyacrylamide/acrylamido
methyl propanesulfonate (AMPS) copolymer and 0.54%
sodium silicate which minimized settling of the
slurry.

Cement C   Neat cement slurry which contains 44.32% fresh water
and 0.4% lignosulfonate retarder.

Cement D   The same composition as Cement A except that the
polyvinyl acetate polymer was not prehydrated prior
to addition of cement solids.

4 Results and discussion

The cement slurries investigated in this work exhibited viscoelastic
properties. The evidence for viscoelasticity in these cement slurries
is shown in the frequency sweep and strain sweep data obtained with
the oscillating rheometer. The results shown in Figures 1 through 4
consists the measurement of cement slurry properties at 0.69 MPa of
pressure at 27^0C. The results shown in Figures 5 and 6 cover
frequency sweeps measured at 2.76 MPa at 67^0C.

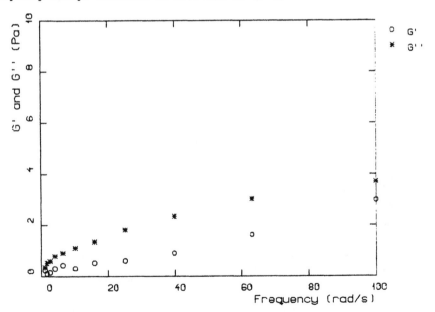

Figure 1.   Frequency sweep of Cement A at 0.69 MPa.

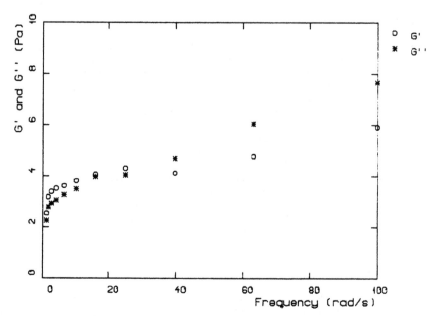

Figure 2.   Frequency sweep of Cement B at 0.69 MPa.

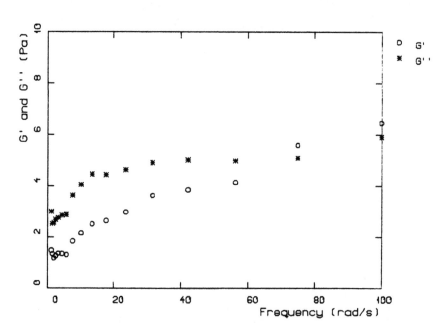

Figure 3.   Frequency sweep of Cement C at 0.69 MPa.

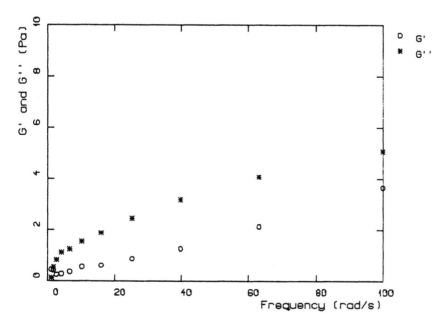

Figure 4.   Frequency sweep of Cement D at 0.69 MPa.

Figure 1 shows frequency sweep curves of Cement A. This slurry has
a recognizable, albeit small, storage modulus (G') which demonstrates
viscoelasticity.   Traditionally, a yield point is calculated using
the 600 and 300 rpm Fann 35 viscometer readings. This indicates a
yield point of 13 Pa (27 lb/100ft^2) for this slurry. However, the
strain sweep data for this slurry as shown in Table 3 indicate that
the yield value is zero. G' is decreasing with increasing strain.
This indicates a continous breakdown of fluid structures, and hence a
flowing state of the fluid.   It could be argued from the low shear
rate data in Table 1, that the yield point is zero.
Figure 2 show frequency sweep results of the cement slurry
containing a polyacrylamide/AMPS copolymer and sodium silicate
(Cement B). G' is larger than G'' for frequencies up to 25 rad/s. One
would expect this to be a gelled fluid. The strain sweep, at low

Table 3. Strain sweep of cement slurries A and B

	Cement A	Cement B
	G'	G'
Strain (% of annular width)	(Pa)	(Pa)
10	1.21	4.99
35	0.50	5.06
60	0.48	7.92

175

strain, is a further indication of a gelled fluid, as G' does not
decrease with increasing strain (Table 3, Cement B). This is also
seen in the Fann 35 viscometer data in Table 1.

The neat cement slurry (Cement C) also exhibits viscoelastic
behaviour (Figure 3). The G' is actually larger in this slurry than
in Cement A which has the added PVA. The elasticity found in neat
cement is due to hydrating cement particle interactions. After the
addition of water, the particles tend to aggregate which results in
higher viscosities. This tendency is affected by particle size,
surface area, water content and mixing. A variety of polymers, such
as the PVA, are capable of adsorbing onto the particle surfaces and
inhibit aggregation, resulting in a thinner slurry. Thus, these
polymers may act like a weak dispersant. Hence, a reduction in the
storage modulus G' may be expected.

Figure 4 shows the storage and loss moduli of Cement D which is a
slurry similar to Cement A. The difference is that the PVA is not
prehydrated, giving less adsorption onto the cement particle
surfaces. Thus, the values lie between those of the neat slurry
(Cement C) and the prehydrated PVA slurry (Cement A).

The measurements at 2.76 MPa at 67^0C show increased values for
both G' and G''. Figure 5 shows the frequency sweep data for Cement
B. G' and G'' have magnitudes approximately five times greater than
the lower pressure test values. Viscoelasticity is clearly
demonstrated by the G' values. In this case G'' is always larger
than G'. The linearity in the G'' values at low frequency indicates
a relatively constant viscosity. The curve implies that this slurry
has a yield value. The decrease of both the G' and G'' values at
the higher frequencies is assumed to be a result of non-linear

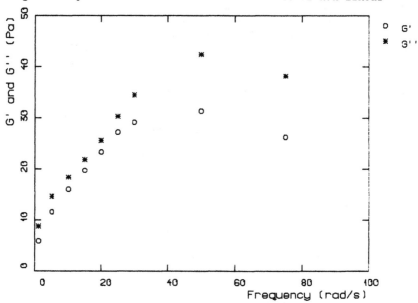

Figure 5. Frequency sweep of Cement B at 2.76 MPa.

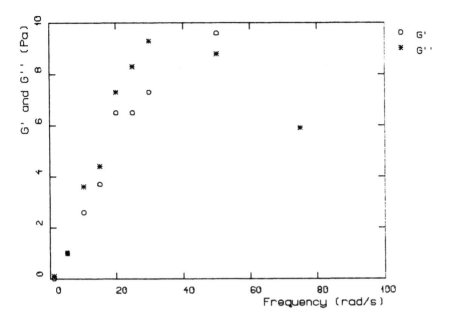

Figure 6. Frequency sweep of Cement C at 2.76 MPa.

effects.

The data for Cement C, as shown in Figure 6, suggest similar trends. In this case, the magnitude of the G' and G'' values are about two times greater than for the lower pressure tests. The measurements suggest a zero yield value for this slurry. Therefore, under these conditions, this slurry may be classified as a Boger fluid; that is an apparently Newtonian fluid with viscoelastic behaviour.

The increased G' and G'' values in these latter tests could be the result of chemical changes such as curing reactions.

5 Conclusions

Viscoelastic properties have been observed in class G cement slurries which have their application in the oil industry. The viscoelasticity was not restricted to gelled slurries or slurries with yield values. The viscoelastic properties can be used to determine whether a slurry has a yield value.

6 Acknowledgements

The authors would like to thank The Royal Norwegian Council for Scientific and Industrial Research (NTNF) and BJ Services (BJ Titan) for their support of this study.

7 References

American Petroleum Institute (1986) Specifications for Materials and Testing for Well Cements, API Specification 10, Third edition. American Petroleum Institute, Dallas, Texas.

Bird, R.B., Armstrong, R.C. and Hassager, O. (1987) Dynamics of Polymeric Liquids, Volume 1, Second edition. John Wiley & Sons, New York.

Chow, T.W., McIntire, L.V., Kunze, K.R. and Cooke, C.E. (1988) The Rheological Properties of Cement Slurries: Effects of Vibration, Hydration Conditions, and Additives. SPE Production Engineering, 3, 543-550.

Cooke, Jr., C.E., Gonzalez, O.J. and Broussard, D.J. (1988) Primary Cementing Improvement by Casing Vibration During Cement Curing Time. SPE Production Engineering, 3, 339-345.

PART FOUR

PROPERTIES OF GROUTS AND GROUTING PROCESSES

9 THE RELATIONSHIP BETWEEN EARLY AGE PROPERTY MEASUREMENTS ON CEMENT PASTES

P.L. DOMONE and H. THURAIRATNAM
Department of Civil and Environmental Engineering,
University College London, UK

Abstract
The development of the early age properties of cement pastes have been measured by a range of tests, including helical impeller viscometry, penetration resistance, Vicat setting, isothermal calorimetry and ultrasonic pulse velocity. Ordinary Portland and High Alumina cement grouts were tested, at water-cement ratios between 0.3 and 0.6. For each test a characteristic time representing a stage in the hydration process was defined. The effect of the mix variables on the various characteristic times showed a consistent pattern, and linear correlations between the times were excellent. The most significant test result in terms of loss of fluidity is the viscometer delay time. However, ultrasonic pulse velocity has the advantage that the results can be used to estimate loss of fluidity, initial set and alite heat of hydration peak from correlations presented in this paper, but also continued monitoring throughout the hardening period can give estimates of strength development.
Keywords: Ordinary Portland and High Alumina cement paste, setting, hardening, rheology, penetration resistance, calorimetry, ultrasonics.

1 Introduction

The "early age" properties of a cement paste, mortar or concrete can be generally defined as those relating to any part of the period from shortly after mixing i.e. when the fresh properties, such as workability, begin to alter, to the establishment of full or nearly full strength, say at 28 days. For most purposes, the three most important parts of the behaviour are:

a) the rate of loss of fluidity or workability, to the point of the so-called initial set, which represents the end of the usable life of the fresh mix;
b) the onset and subsequent rate of strength gain, when loads or stresses can be carried;
c) the period of thermal movement due to heat of hydration effects.

The investigation reported in this paper concentrates mainly on comparing tests used to assess the first of these - the loss of fluidity and setting – of cement paste or grout, with some extension into the period of onset of strength gain.

The work has formed part of a more comprehensive study of the properties of cementitious grouts used in offshore structures; although aimed at offshore practice, in many respects the materials and mixes used do not differ from those used for more widespread purposes, and therefore the data has direct relevance to the monitoring and understanding of the behaviour of pastes and grouts in general.

2 Materials and mixes

Three cements were used, Ordinary Portland, Oilwell B and High Alumina. All could be considered typical of their type; their composition and major physical properties are given in Table 1.

For each cement, mixes of water-cement ratios 0.3, 0.4 and 0.5 were tested, with 0.6 also tested for the OPC. For all three cements, fresh (tap) water was used for mixing, with sea water also being used for the OPC mixes; this was mixed in the laboratory to have the composition given in Table 2. Mixing and curing was normally carried out at 20°C, but the effect of a lower temperature (8°C) was also assessed on the fresh water/OPC mixes.

Table 1. Analyses of cements.

	OPC	Oilwell	HAC
Oxides (% by weight)			
SiO_2	20.1	21.2	4.65
Al_2O_3	5.1	3.5	38.88
Fe_2O_3	3.4	5.4	10.9
CaO	64.0	65.0	37.51
MgO	1.3	0.6	trace
SO_3	3.0	2.1	0.17
K_2O	0.80	0.22	0.11
Na_2O	0.13	0.10	trace
I.R.	0.85	0.19	
L.O.I.	0.9	1.1	0.81
free CaO	0.8	1.0	
FeO			4.85
Bogue compound composition			
C_3S	56.8	62.2	
C_2S	14.8	13.9	
C_3A	7.8	0.14	
C_4AF	10.3	16.4	
specific surface area (m^2/kg)			
	351	333	300

Table 2. Composition of artificial sea water.

Salt	% by wt	ion	% by wt
NaCl	2.80	Na^+	1.100
$MgCl_2$	0.34	Ca^{++}	0.044
$MgSO_4$	0.20	Mg^{++}	0.126
$CaSO_4$	0.15	Cl^-	1.953
		SO_3^{--}	0.222

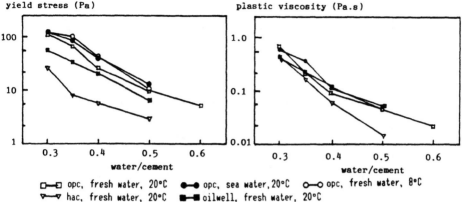

yield stress (Pa)   plastic viscosity (Pa.s)

water/cement   water/cement

□—□ opc, fresh water, 20°C ●—● opc, sea water,20°C o—o opc, fresh water, 8°C
▽—▽ hac, fresh water, 20°C ■—■ oilwell, fresh water, 20°C

Fig. 1 Bingham properties of the pastes.

Mixing was carried out using a high shear Silverson disintegrating head mixer, shown previously to give very efficient dispersion of the cement i.e. sufficient to give reversible flow curves in a concentric cylinder viscometer, Domone and Thurairatnam (1988).

3 Testing

Immediately after mixing the pastes were checked for initial fluidity using a conventional concentric cylinder viscometer with a bob of radius of 12.5mm and length 37.5mm and a cup of radius 13.55mm. This was mounted in a Contraves Rheomat/Rheoscan control and measuring system, described previously, Domone and Thurairatnam (1988). The Bingham properties of yield stress and plastic viscosity were calculated from the flow curves, and these were found to correspond to previous results on similar mixes (fig 1).

The early age behaviour, in effect the progress of hydration, of each mix was then assessed by a number of tests, which were all relatively well established before the start of the work. For each test a "characteristic time", representing either a significant stage in the hydration or a distinct change in behaviour was defined. The initial aim was to to use the tests to characterise the mixes, but as will be seen an excellent linear correlation between the results was obtained.

The tests and the definition of the characteristic times were as follows.

3.1 Penetration Resistance.

A 6mm diameter plunger mounted on a drill stand was lowered into the paste, and the force needed for a 25mm penetration measured with a spring balance. The test is fully described in ASTM 403-77 (1977). A penetration pressure of $3.5N/mm^2$, corresponding to a force of 10.1kg, is normally taken to be the initial set, i.e. the end of the usable fluid life of the grout. The time taken to reach this value is referred to as PR_i.

Typical results for the OPC pastes are shown in fig 2.

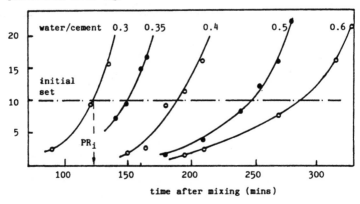

penetration force (kg)

Fig.2. Development of penetration resistance for OPC pastes at 20°C.

3.2 Modified Vicat test.
The initial set was determined using a standard Vicat apparatus, BS
4550 (1977), but testing the range of pastes rather than a mix of
standard consistence. As might be expected, the set times had an
excellent aggreement with those from the penetration resistance tests
for all mixes, and therefore.are not reported in this paper.

3.3 Helical impeller viscometer
The cylindrical bob in the viscometer described above was replaced
with an interrupted helical impeller (fig 3). This was run continuously
at an angular velocity of 42sec^{-1}, the torque increasing with build up
of hydration product. The angle of the blades was set at
45°, and the ratio of cup to blade diameter was 1.16. Such an impeller
has been shown by Bhatty and Banfill (1982) to provide sufficient
mixing action to give negligible sedimentation of the paste.
Calibration against the concentric cylinder system showed that it could
be used for Bingham property determination, but in these tests it was
only the change in torque, and not the interpretration of its value in
fundamental terms, that was of interest.
    It was found, in common with Bhatty and Banfill (1984), that there
was a well defined time at which a distinct increase in slope in the
log(torque) vs log(time) graph occurred (fig 4). This indicates a more
rapid build up of hydration product after this time, which was taken as
the characteristic time for this test. It has been termed "delay time"
(t_d) by Bhatty and Banfill, and they have suggested that it could
represent the end of the dormant period of cement hydration.

3.4 Constant temperature calorimeter
The rate of heat output during hydration at constant temperature was
measured in a Wexham type calorimeter, Forrester (1970), using a 50gm
sample of the paste, The results were expressed as heat output per unit
weight of cement. The sample was loaded into the cell after mixing, and
so the first peak recorded is in fact the second or alite peak. The
characteristic time (referred to as CAL₁) was taken as the time to
reach this peak; typical results are shown in fig 5.

184

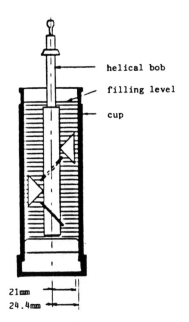

21mm
24.4mm

Fig. 3. The interrupted helical
impeller system.

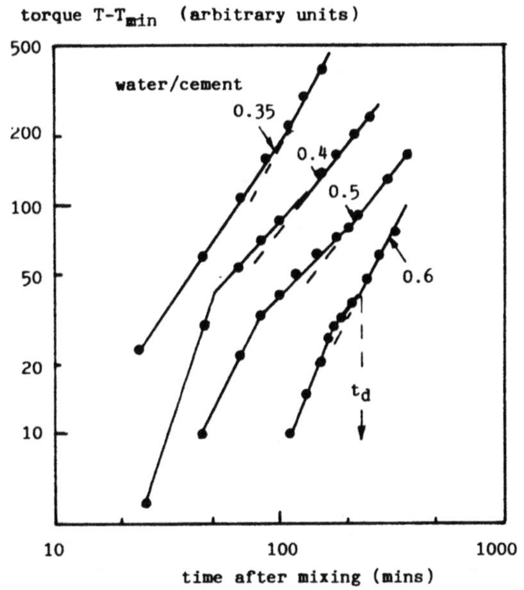

torque $T-T_{min}$ (arbitrary units)

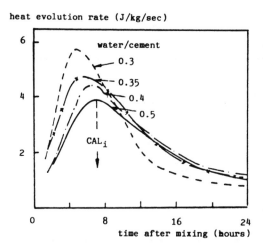

500

water/cement

200   0.35

0.4

100   0.5

50   0.6

20   t_d

10

10   100   1000

time after mixing (mins)

Fig. 4. Development of torque with time
in helical impeller viscometer
for OPC paste at 20°C.

heat evolution rate (J/kg/sec)

6

water/cement
0.3
0.35
4   0.4
0.5

CAL_i
2

0   8   16   24

time after mixing (hours)

Fig. 5. Rate of heat output of OPC paste at 20°C measured in constant
temperature calorimeter.

185

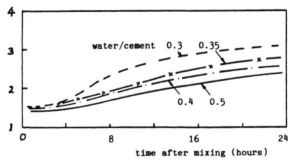

ultrasonic pulse velocity (km/sec)

water/cement 0.3 0.35

0.4 0.5

time after mixing (hours)

Fig.6. Development of ultrasonic pulse velocity with time for OPC pastes
at 20°C.

3.5 Ultrasonic pulse velocity

The transit time of an ultrasonic pulse through a 30mm thick sample of
the paste was measured with a Pundit apparatus linked to a data
logger. The paste was contained in a polythene bag, which was placed
between two waterproofed 54kHz transducers mounted in a PTFE support
block. The whole system was placed under water, which ensured good
contact between the transducers and the bag of paste, and uniform
temperature. Apart from the data logging, the test system was as
previously described, Casson and Domone (1982), and has been
extensively used for monitoring both mortars and cement pastes.

The transit time was converted to pulse velocity, and, using a
similar approach to heat of hydration, the rate of change of pulse
velocity was then calculated. This was found to occur consistently at a
pulse velocity of about 1.75km/sec, and therefore the time to reach
this velocity was adopted as the characteristic time (figs 6 and 7).

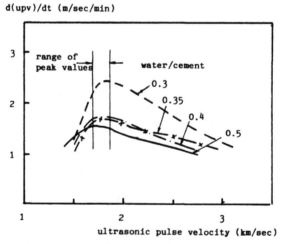

d(upv)/dt (m/sec/min)

range of
peak values

water/cement

0.3

0.35

0.4

0.5

ultrasonic pulse velocity (km/sec)

Fig.7. First derivative of pulse velocity with time, d(upv)/dt, plotted
against pulse velocity; derived from data in fig.6.

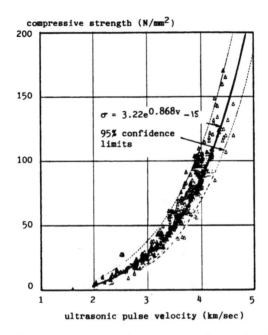

compressive strength (N/mm²)

ultrasonic pulse velocity (km/sec)

Fig 8. Compressive strength vs ultrasonic pulse velocity for hardened
cement paste (data for a variety of mixes and ages).

In the course of concurrent studies at UCL on the rate of strength
gain, relationships between ultrasonic pulse velocity and compressive
strength were established, Domone and Thurairatnam (1986, 1988). From
results for more than 500 cubes, made from several types of cement,
with and without admixtures and tested at ages from 12 hours to a year,
a best fit correlation of:

$$\sigma = 3.22e^{0.868v} - 15$$

where σ = cube compressive strength in N/mm²
and v = ultrasonic pulse velocity in km/sec,
was obtained (fig 8). Extrapolating this expression to σ = 0, gives v =
1.77km/sec, very close to the value of 1.75km/sec chosen for the other
reasons described above. It would therefore seem that this has some
significance in terms of microstructure development, perhaps the onset
of true strength gain.

Results and discussion

Figures 9, 10, 11 and 12 show the variation of the characteristic times
PR_i, t_d, CAL_i and UPV_i with water-cement ratio for each of the mix. t_d
is generally the shortest time, closely followed by PR_i, then CAL_i
sometime later. UPV_i is very similar to CAL_i at low water-cement
ratios, but longer at higher values. For each test, there is an
increase of characteristic time with water-cement ratio, and the
general pattern of behaviour is similar with each test:

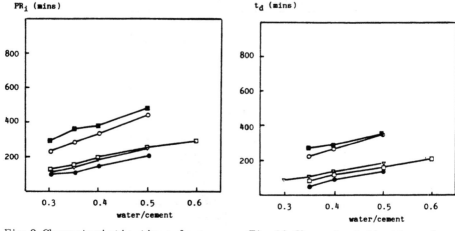

Fig.9.Characteristic times from
penetration resistance (PR$_i$).

Fig.10.Characteristic times from
helical impeller viscometer (t$_d$).

□—□ opc, fresh water, 20°C   ●—● opc, sea water,20°C   ○—○ opc, fresh water, 8°C
▽—▽ hac, fresh water, 20°C   ■—■ oilwell, fresh water, 20°C

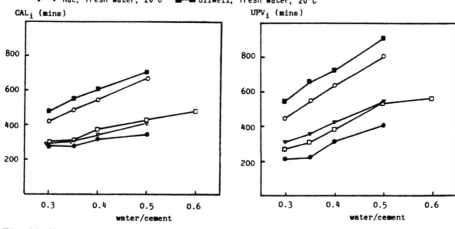

Fig.11.Characteristic times from
calorimeter (CAL$_i$).

Fig.12.Characteristic times from
upv (UPV$_i$).

- the Oilwell cement mixes have considerably longer characteristic
  times than the equivalent OPC mixes;
- the lower temperature (8°C) causes considerable extension of the
  characteristic times for the OPC mixes, bringing them close to the
  times for the Oilwell cement mixes at 20°C.
- sea water acts as a low efficiency accelerator;
- the HAC mixes have longer values of t$_d$ than the equivalent OPC
  mixes, indicating a longer workable life and so-called "dormant"
  period, but the PR$_i$ values are similar and the CAL$_i$ values shorter,
  showing that the rapid increase in rate of hydration after the
  dormant period quickly dominates the behaviour. The UPV$_i$ values are
  a little longer than those of the OPC mixes.

All the tests measure a different property of the paste, and therefore represent a different stage in the hydration process. For example as mentioned above, t_d is probably related to the end of the dormant period, and UPV_i could represent the onset of measurable strength, although in limited previous tests on mortar, the peak of the $d(upv)/dt$ curves appear before any significant microstructure could be discerned with an electron microscope, Casson et al (1982). Furthermore, in at least one case (PR_i) the characteristic time is somewhat arbitrarily defined.

It was therefore not necessarily to be expected that well defined correlations between the results would exist. However, it was in fact found that there were excellent, linear correlations for the range of mixes tested. The torque delay time (t_d) possibly represents the most fundamental property in terms of change of rheology (and also occurs first), and the other times are shown plotted against this in fig 13. For each test, the results for all of the mixes all fall close to the same straight line, with possibly a distinction between the Portland cement mixes (i.e. those with OPC and Oilwell cement) and the HAC mixes for the PR_i and CAL_i results. These two sets of results are approximately parallel, with the UPV_i vs t_d having a steeper slope. A similar linear correlation between initial Vicat setting time and t_d was obtained by Bhatty and Banfill (1984).

The best fit linear regression equations and the correlation coefficients are given on the graphs. If extrapolated none would go through the origin.

Equivalent linear correlations of t_d, PR_i and CAL_i with UPV_i could equally well be drawn, and the regression equations are also given in fig 13. The ultrasonic pulse velocity test does, however, have the major advantage that it can continuously monitor the paste well into its hardened state, and therefore can be used to estimate strength development (using such data as shown in fig 8) and other mechanical properties such as elastic modulus. It is relatively simple test to carry out and to link to automatic data logging and analysis, and therefore it is suggested that, if convenient correlations such as those given apply for a greater range of variables (admixtures, cement replacement materials etc), then this is an extremely useful test for estimating a significant range of properties.

Subsequent to the commencement of work reported in this paper, further tests for assessing early age properties have been suggested, most notably the measurement of shear modulus at very low shear displacements, Keating et al (1989), and the measurement of electrical properties of dielectric constant and resistivity, McCartur et al (1988). The shear modulus tests were carried out in parallel with upv measurements using a similar system to ourselves, the only significant differences being the use of higher frequency transducers (200kHz) at a longer path length (40mm). It was found that the shear modulus started to increase during the first one or two hours after mixing i.e. before the upv registered any significant change. The electric properties appeared to show changes at about the same time as our upv results. These latter tests would apear to have the same advantages as upv in being able to monitor the specimens through into the hardened state. Some further comparative tests would perhaps be useful in assessing the relative merits of each of techniques.

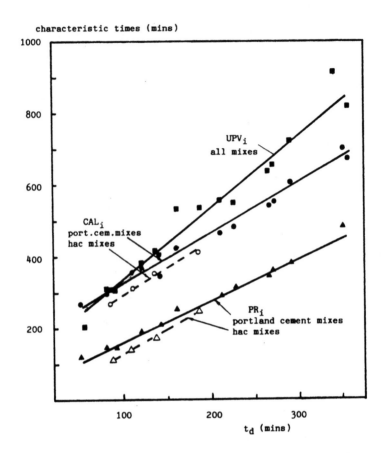

characteristic times (mins)

regression equations	corr.coeff	regression equations	corr.coeff
$UPV_i = 141 + 1.99t_d$	0.983	$CAL_i = 114 + 0.66UPV_i$	0.992
$CAL_i = 194 + 1.37t_d$	0.992	$PR_i = -23 + 0.56UPV_i$	0.996
$PR_i = 52 + 1.14t_d$	0.994	$t_d = -62 + 0.48UPV_i$	0.983

Fig.13.Correlations between UPVᵢ, CALᵢ and PRᵢ with tᵈ.

5 Conclusions

The results have indicated that the change in torque required to drive
a helical impeller at constant speed in a viscometer can be used to
provide useful information on the progress of hydration in a cement
paste. "Characteristic" times for this and a number of other tests used
to assess the build up of microstructure can be defined, and although
these times all represent different, and sometimes abitrary, stages in
the hydration process, linear correlations between them were obtained.
It would seem that ultrasonic pulse velocity is the most versatile in
assessing the setting, hardening and early strength gain processes.

Acknowledgements

The work reported in this paper was part of the Managed Programme on
Grouts & Grouting for Offshore Structures which was promoted by the
Marine Technology Directorate and sponsored jointly by SERC, Department
of Energy and the offhsore industry. The authors would also like to
acknowledge the assistance in the experimental programme of the
technical staff of the Elvery Concrete Technology Laboratory at
University College London.

References

American Society for testing and Materials "Standard test method for
    the time of setting of concrete mixtures by penetration resistance"
    ASTM C 403 - 77, 1977
Bhatty J I and Banfill P F G "Sedimentation behaviour in cement pastes
    subjected to continuous shear in rotational viscometers" Cement and
    Concrete Research, Vol 12, No 1, 1982 pp 69-78
Bhatty J I and Banfill P F G "A viscometric method of monitoring the
    effect of retarding admixtures on the setting of cment pastes"
    Cement and Concrete Research, Vol 14, No 1, 1984 pp 64-72
British Standards Institution BS 4550: Part3: Section 3.6 Tests for
    setting times of cement" 1977.
Casson R B J and Domone P L "Ultrasonic monitoring of the early age
    properties of concrete" Proc of RILEM International Conference on
    "Concrete at Early Ages" Paris, April 1982, pp 129-135
Casson R B J, Domone P L, Scrivener K, Jennings H M, Gillham C J and
    Pratt P L "The use of ultrasonic pulse velocity,penetration
    resistance and electron microscopy to study the rheology of fresh
    concrete" Materials Research Seminar, Boston, USA, November 1982.
Domone P L and Thurairatnam H "Development of mechanical properties of
    of ordinary Portland and Oilwell B cement grouts." Magazine of
    Concrete Research, Vol 38, No 136, Seotember 1986 pp 129-138
Domone P L and Thurairatnam H "The effect of water/cement ratio,
    plasticizers and temperature on the rheology of cement grouts."
    Advances in Cement Research, Vol 1, No 4, October 1988 pp 195-206
Domone P L and Thurairatnam H "Cement grouts with specific properties
    for offshore use" in "Grouts and grouting for construction and
    repair of offshore structures - a summary report" Department of
    Energy Offshore Report OTH 88-289, 1988 pp 44-58.
Forester J A "A conduction calorimeter for the study of cement
    hydration" Cement Technology, May/June 1970 pp 95-99.
Keating J, Hannant D J and Hibbert A P "Comparison of shear modulus and
    pulse velocity techniques to measure the build-up of structure in
    fresh cement pastes used in Oilwell cementing" Cement and Concrete
    Research Vol 19 No 4 July 1989 pp 554-566
McCarter W J and Afshar A B "Monitioring the early hydration mechanisms
    of hydraulic cement" Journal of Materials Science, Vol 23, 1988
    pp 488-496

20 PUMPING OF CEMENT GROUTS

S.A. JEFFERIS
Queen Mary and Westfield College, London, UK

Abstract
This paper is concerned with the pumping of cementitious grouts and
particularly the development of a lubricating layer at the pipe wall
during flow. The effect of flow velocity and pipe diameter on layer
thickness are deduced from analysis of data from many different
pumping tests carried out over a period of several years. It is
found that the data are susceptible to many different
interpretations. One interpretation is that for low velocities and
small pipes layer thickness is independent of flow velocity and
inversely proportional to the pipe diameter. Thus it is possible
that a classical approach based on the ratio of particle diameter to
pipe diameter may have some application.
    A fuller understanding of the lubricating layer may enable better
predictions of pumping pressures from empirical tests such as flow
cones particularly if the role of the lubricating layer in these
tests can be assessed. In practice this may not be possible but it
may be that test procedures can be screened for their relevance to
the pipe flow situation.
Keywords: Grouts, Pumping, Lubricating layer, Cement.

1 Introduction

Before considering the problems involved in the analysis of grout
pumping it is appropriate to ask what is the problem? Industries
which are involved in grout pumping include: mining for backfilling,
formation sealing and roadway stabilisation; oil well drilling for
the cementation of conductor pipes to the formation, the offshore oil
industry for structural connections and the construction industry for
the filling of pre-stressing ducts to protect the tendons. Also many
repairs grouts are now pumped to enable rapid and accurate placing.
    In general it would seem that rather few problems have been
reported with the pumping of grouts. Those that do occur are often
related to blockage of lines by jamming with oversize solids
(hardened grout dislodged from mixers etc.) or from the loss of water
at joints etc. However, most of those involved in grouting would
allow that better data on pumping pressures could bring savings in
the power provisions for the pumping equipment. Other problems
relate to the prediction of pumpable time. If a hold up occurs can

grout be safely left in the pipe for a defined time or should it be dumped immediately? In all the industries there is also the problem of the penetration of the grout into the interstices of the zone to be grouted. In many situations this penetration will be by hydrostatic pressure of the grout caused by differences in surface level rather than the applied pumping pressure. Thus problems that need to be addressed are: peak pumping pressures, pumpable time and flow under small hydrostatic pressures.

From the research standpoint rheological investigations of grouts are widely used to identify microstructural effects of hydration, admixtures etc. However, for this rotational viscometry is much more convenient than measurement of pumping behaviour. Though it must be accepted that the two procedures will not give the same results! Particularly as the residence time of the grout in a pipeline may be from a few seconds to perhaps half an hour and is controlled by the flow rate and length of pipe whereas in a rotational viscometer the residence time is under the control of the operator within the limits set by stiffening of the material.

2 Microstructure of grout

To analyse the behaviour of grout it is necessary to consider the nature of the material. When a grout is mixed with water the cement grains will tend form a coarse dispersion in the water phase. The solid phase may not be discrete grains but aggregates of many grains. These aggregates may be broken down by shear during mixing. Thus the grading of the particles in the suspension may depend on the shear level and time of mixing and the time after first contact with water that the mix is subjected to shear as this will influence the strength of the bonding within the aggregates. This may be particularly significant for accelerated mixes, Jefferis (1985).

When first mixed a grout will behave as a two phase material. However, hydrate will be forming on the surface of the cement grains and may be partially dispersed into the aqueous phase as a colloidal material. Thus with time and mixing the aqueous phase will tend to become a viscous colloidal fluid. As a result the overall system will tend to move towards a homogeneous material as the solid particles become progressively more bound up in the viscous aqueous phase so that the separation of the solid and liquid phases becomes more difficult. Therefore the rheology of a grout may be substantially influenced by the type of mixer employed and the mixing time. For example before the advent of specialist admixtures high shear mixing was used to produce cohesive grouts for placement through water.

Figure 1 shows data for Shear stress at the wall as a function of time in a recirculating pumping rig for a high shear mixed grout and a low shear mixed grout of the same water/cement ratio. It can be seen that the plot for the high shear mixed grout (prepared in a commercial high shear grout mixer) shows a steady increase with time. It has been found that the rate of increase may be limited by cooling the grout and thus the rise with time would seem to be related to hydration effects. In contrast the grout prepared by low shear

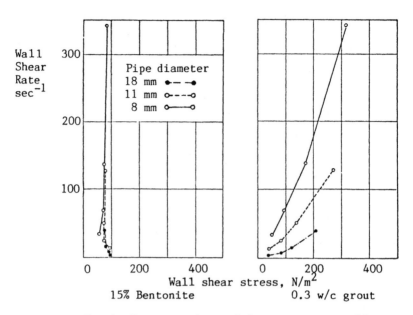

Fig.1. Specimen plots of Consistency variables.

mixing in a planetary motion paddle mixer shows a rapid drop in the first minutes of pumping and then a steady decline to a plateau which may be held for an hour or more (depending on the mix and again the temperature) and then a slow increase. Very often in tests with low shear mixed grouts it is found that the grout becomes unpumpable (it will not pass through the pump, see Section 3) before the pumping pressure returns to or exceeds its initial value. Thus for these grouts it seems that the effects of hydration are manifested much more slowly. With low shear mixed grouts it is often difficult to record the initial peak. Thus for analysis data from the plateau region may be used. For design such figures may need to be increased by up to 90% to allow for the peak pressure. The cause of the initial peak and subsequent reduction is not obvious. The reduction is not solely due to mixing of the grout by the pump (breaking down of coarse lumps) as the cycle time for an element of grout through the loop was of order 1 minute and the drop started before this. Thus either breakdown was occurring within the pipe or the reduction was from some other cause such as the rather slow development of a lubricating layer. A thicker lubricating layer would be expected from a low shear mix than a high shear mix as the aqueous phase would be more fluid and thus separation would be easier.

3 Water/cement ratio

Grouts can be used over a very wide range of water/cement ratios depending on the industry and the application. In construction the range may be from perhaps 0.25 to 0.6. The porosity of such mixes will be of order 0.44 to .65 and thus if the cement were not reactive it could be expected that the suspensions would be very loose. That is that the particles would settle on standing to leave a significant volume of free water. Also it could be expected that even the drier mixes would flow without shear thickening due to dilatancy effects. However, in practice it is found that rather little settlement occurs except at the higher end of the water/cement ratio range (though this will be very dependent on the shear rate during mixing and the geometry of the sample). Furthermore dilatancy effects are observed when pumping low water/cement ratio mixes especially after significant pumping times. Indeed in laboratory trials it has been found that the time for which a mix is pumpable cannot be estimated from pumping pressure - time plots. Rather pumpability is controlled by the ability of the grout to pass through the pump where it is subjected to large strains and thus dilatancy effects can be severe.

4 Phase separation during flow

As a result of the internal shear during flow there must be expected to be a tendency for separation of the phases. Such separation will be inhibited by the developing structure which links particles and limits the net compression of the matrix and expulsion of the aqueous phase. Also the viscosity of the aqueous phase could be expected to limit the rate of separation and hence in a dynamic equilibrium the amount of separation. During flow two types of separation could be proposed: preferential migration of water or preferential migration of solids to the boundaries. Migration of water to the wall would reduce pumping pressure whereas migration of solids would tend to increase pumping pressures. Would such a movement be contrary to the spirit of Le Chatelier's principle? Could a pumped grout be regarded as a system at equilibrium to which a constraint has been applied and thus one which would be expected to move so as to minimise the constraint (the pumping pressure)? Is the movement of water so as to reduce pumping pressures thus thermodynamically predictable?

5 Pressure on the pipe wall

If an attempt were made to push an air/cement mix along a pipe the pumping pressure would be very high due to the friction of the grains at the wall. The system would very probably block since increased pressures would increase friction forces and so on. In such a system the pressure in the air phase could be much lower than the average stress in the solid phase. Thus in principle the radial stresses on the wall of a pipe may have two components (a) the pressure from the solid phase (this might be expected when pumping harsh mixes) and (b) the pore pressure of the fluid phase.

To investigate the possible existence of solid friction a series of tests were undertaken in which the total radial stress of the grout on the pipe wall was measured and also the pore pressure of the liquid phase. In principle the procedure was very simple. Grout was pumped along a smooth steel pipe. At intervals along the pipe there were short sections that had been machined so as to leave a wall thickness of about 0.5 mm. The thinned sections were fitted with strain gauges so that the internal pressure could be measured. Close to each of these sections there was a pressure tapping connected to a pressure transducer via a porous stone set at the inner wall of the pipe. In this way the strain gauge system measured the total stress on the pipe and the pressure transducers the pore fluid pressure.

In practice there were many problems with the system. Zero readings had to be taken with an empty pipe and there was a temperature change when the grout was introduced. This temperature change caused significant strains and although the effects were minimised by the use of gauges temperature matched to the steel of the pipe the errors were unacceptable. It was found necessary to actually measure the temperature and apply a correction obtained from prior calibration of the gauges. In this way temperature effects were much reduced but not eliminated and they reduced the sensitivity of the system. There were also problems of bending of the pipe when pumping started. This was partially eliminated by the use of a flexible coupling between pump and pipe.

In a series of tests involving relatively dry grouts of water/cement ratio 0.275 to 0.375 it was found that there was no measurable difference between the pore pressure measured with the transducers and the and total stress measured with the strain gauges. It would therefore seem that there was no significant solid pressure on the wall of the pipe for these grouts and thus for any grouts of higher water/cement ratio. However, when pumping harsh concretes it would seem that there could be effects from solid contact stresses.

6 Lubricating layers

Of course it is generally accepted that the material in contact with the wall of a pipe may be more fluid than the bulk material when pumping grouts or concrete. However, it was felt necessary to undertake the tests described above so as to confirm that this concept is not totally ill founded and that pressure losses are not grossly influenced by the friction of solids.

Various mechanisms can be identified which would tend to drive the solids away from the wall and thus make it a region rich in fluid. In any pipe flow situation the wall region will be the region of highest velocity gradient. Solid particles in this velocity field may be subjected to radial inward lift forces. Also dilatancy effects in a region of high shear may tend to separate the particles. Thus the flow regime in the pipe might be as follows:

(a) a wall region of higher water/solids ratio than the input material. The rheological properties of this layer cannot be predicted save that it will be more fluid than the core region. The thickness of the layer also is unknown.
(b) a core region of lower water/solids ratio than that of the input material.

In principle if the water/solids ratio of the wall layer were known
(or strictly the distribution of water/solids as the layer is
unlikely to be of uniform concentration throughout its thickness)
then a mass balance could be carried out to determine the thickness.
However, allowance would have to be made for the different average
velocities of the wall and core regions.

In practice it has been found extremely difficult to get any
information on the thickness or rheology of the wall layer.
Theoretical considerations would suggest the following:

(a) Severe turbulence in the pipe will tend to disrupt the wall
layer. Thus the wall layer may be less significant at high flow
velocities or in large diameter pipes (that is situations of high
Reynolds' number).

(b) The wall and core regions will be rheologically distinct
regions. The thickness of the wall layer cannot be obtained from
application of the Buckingham Reiner equation (the standard
equation for flow of a Bingham fluid) using the properties of the
core material.

Thus the simplest model for the lubricating layer is to assume that
it is a Bingham fluid with properties such that it is all sheared and
there is no plug flow region within the layer. This would seem
reasonable if the origin of the lubricating layer is a velocity
gradient driven separation of water and solids. The core region
inside the lubricating would then be a region of entirely plug flow
with no tendency for velocity gradient induced water migration. The
situation is thus very similar to the conventional sheared region and
plug flow of a Bingham material save that the thickness of the
sheared region is not controlled by the yield stress of the material
in the core region.

The above situation can be analysed and if the wall layer is
assumed to be small in thickness compared with the radius of the pipe
then it can be shown that:

$$P = v \eta / a + \tau \tag{1}$$

where: v is average velocity in the pipe (flow per unit area),
        a is the thickness of the wall layer,
        η is the plastic viscosity of the fluid in the wall layer,
        τ is the yield stress of the fluid in the wall layer,
        P is the stress consistency variable for pipe flow (p.d/4)
        p is the pressure drop per unit length
        d is the diameter of pipe.

It would be more usual to use both stress and shear rate (8v/d)
consistency variables in an equation such as (1) above. However it
has been found that such a formulation is not helpful for pipe flow
of cementitious materials. Figure 2 shows a typical plot of the
consistency variables for the flow of a bentonite clay suspension and
a cement grout. It can be seen that for the bentonite the results
for three different pipe diameters all lie on effectively a single
curve despite the fact that this curve has a most unusual shape due

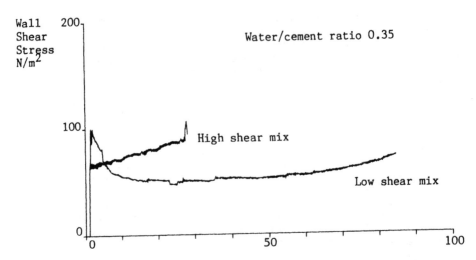

Fig.2. Typical Shear stress – time plots for high and low shear mixes

to thixotropy. However, for the grout the results for each pipe are
quite distinct. A lubricating layer will produce such an effect
though so could other features of the fluid. A simple Bingham fluid
would show a unique consistency curve for all pipes.

For simplicity the thickness of the wall layer could be assumed to
be a function of average velocity in the pipe (total flow rate /
cross-sectional area) and the pipe diameter. Thus:

$$a = b \; v^x \; d^y \tag{2}$$

From consideration of Reynolds' number one would expect that x would
be negative. That is higher velocities would give rise to lower
layer thicknesses though it must be noted that the wall layer may not
form unless there is some shear in the wall region. Similarly y
would be expected to be negative. Combining equations (1) and (2)
and rearranging gives:

$$P = \tau + v^{1-x} \; d^{-y} \; \eta \; / \; b \tag{3}$$

7 Examination of pumping pressure data

Equation (3) was used to investigate the results from some 130 tests
involving six different pipe diameters and many different flow rates
and grout formulations. Investigation of the data showed that for
small pipes and low velocities best fits were obtained for x in the
range 0.1 to 0.5 and y in the range −0.50 to −0.8. Correlation
coefficients typically were in the range 0.98 to .995. However,

urther investigation of the data showed that the implied values of he lubricating layer thickness and yield stress were absurd. urther investigation showed that a reasonable compromise could be btained with x = 0 and y = -1.

For larger pipes and higher flow velocities these values did not ive a good fit (though the data are rather limited). Best fit was btained with x = -2.5 and y = -1. However, these figures gave nreasonable values for the layer thickness and the yield stress and o better compromise values could be identified. Of course under uch conditions (higher velocity, larger diameter) the lubricating ayer would be expected to be very thin and perhaps of no ignificance. The figures of -2.5 and -1 for x and y should not be aken as having any general validity. However, the increasing value f x with pipe diameter and flow rate means that pressure losses in ractice may be much higher than in small scale laboratory models nless careful attention is paid to the scaling rules. For such ituations an analysis based on the average flow velocity in the pipe ay be more helpful, Jefferis and Mangabhai (1988).

Table 1 gives a summary of the results obtained from the modelling xercise outlined above.

Table 1. Investigation of data from pumping tests

Water/ solids ratio	Number of data points	Coefficient of correlation	slope η / b (Ns/m^4)	Range of velocities (m/s)	Range of pipe diameters (mm)
0.375 opc	16	0.96	31000	0.008 - 0.3	8 - 16
0.35 opc	16	0.94	49000	0.008 - 0.3	8 - 16
0.325 opc	16	0.98	87000	0.008 - 0.3	8 - 16
0.30 opc	16	0.98	119000	0.008 - 0.3	8 - 16
0.275 opc	16	0.99	202000	0.008 - 0.3	8 - 16
0.3 10% pfa*	8	0.99	141000	0.02 - 0.3	8 - 11
0.3 20% pfa	8	0.99	149000	0.02 - 0.3	8 - 11
0.3 30% pfa	8	0.99	147000	0.02 - 0.3	8 - 11
0.3 40% pfa	8	0.99	153000	0.02 - 0.3	8 - 11
0.3 50% pfa	8	0.99	153000	0.02 - 0.3	8 - 11
0.35 opc	4	0.97	15000	1.4 - 5.6	8 - 16
0.4 opc	4	0.99	16000	1.4 - 5.6	8 - 16
0.4 oilwell B	5	0.99	29[+]	2 - 7	19 - 40

* pfa added at 10% of total solids
+ best fit for these results was x = -2.5, y = -1, note: units given at head of column must be adjusted accordingly

It is interesting to note that the slope η/b increases substantially as the water/cement ratio is reduced for the opc grouts at low flow velocities. However, the effect on the wall layer thickness is limited as the viscosity η also could be expected to increase significantly. Using reasonable values for the viscosity gives wall layer thicknesses of order 1 to 2 mm for the 8 mm diameter pipe and 2 to 4 mm for the 16 mm pipe. It seems that η/b is little affected by pfa substitution over the range investigated.

8 Prediction of pumping pressures from other test data

The amount of equipment and labour necessary to obtain reliable pumping pressure data is very substantial. There is thus a continuing interest in the use of small scale tests such as a viscometer, flow cone, flow trough, penetrometer etc. to predict pumping pressures. A priori this may seem reasonable but in practice if a range of grouts is tested it is found that the various tests do not give the same ranking of the grouts as regards fluidity nor do they show the same trends with time from mixing. Thus the reading of a rotational viscometer may change rather little over a period of minutes/hours as the grout is effectively kept stirred. Whereas the time ofbflow from a flow cone may change by more than an order of magnitude. As pumping involves continuous agitation of the fluid it would seem in principle that the best tests for assessment of pumping pressure would be those which involve mechanical agitation of the fluid. However, a simple rotational viscometer may produce conditions in which the lubricating layer exerts an undue influence. Some disturbance of the layer may be important and thus mixer type rheometers (Tattersall and Banfill, 1983) or the consistometer (American Petroleum Institute, 1980) may be more suitable though these will underestimate the effects of the lubricating layer.

9 Acknowledgements

The author gratefully acknowledges the work of the many research students who have contributed to the database on grout pumping first at King's College and later at Queen Mary and Westfield College. In particular this paper draws on the work of Dr. Sakuta, Mr. Mangabhai.
    The author is also pleased to acknowledge the contribution of The Marine Technology Directorate Limited who financed two programmes of pumping work by the author and provided a forum for feedback from the grouting industry.

10 References

American Petroleum Institute (1980) Publication RP10B, Testing oil-well cements and cement additives.
Jefferis, S.A. and Mangabhai, R.J. (1988) Laboratory measurement of grout pumping pressures, in Grouts and grouting for construction and repair of offshore structures, HMSO, pp91-110.
Jefferis, S.A. (1985) Discussion on the Arab potash solar evaporation system, Proc. Instn. Civ. Engrs. Part 1, pp641-646.
Tattersall, G.H. and Banfill, P.F.G. (1983) The rheology of fresh concrete, Pitman Books Ltd., London.

21 POSSIBILITIES OF OPTIMISING THE PROPERTIES OF PRESTRESSING GROUTS

W. MUSZYNSKI and J. MIERZWA
Technical University of Krakow, Poland

Abstract
The proces of injection of the longitudinal cable channels of prestressed constructions is a very liable and from the technological
point of view incovenient treatment. The optimum composition of
suspension must comply with a number of opposing conditions like
liquidity, strength, resisntance to frost or sedimentation limited in
time. Investigation proved that the mathematical description of the
injection flow and sedimentation phenomena is possible basing on the
previously stated characteristics of the pumped cement paste. Since
the dependencies between rheological parameters of cement paste and
its coefficient of solids concentration, together with other
physical and mineral characteristics of cement, are known it turned
out to be possible to reduce an anylysis to the process of
optimizing the objective function which can in general be expressed
as $\omega_{opt} = (w/c)_{opt} = \omega_{min}$ with admitted limitations. In this paper
the above idea was developed and formalized in the algorithmical
shape.
Keywords:Rheology, Injection Grout, Cable Channel, Sedimentation.

1 Introduction

The injection of a cable channel of a prestressed construction is a
very responsible operation as its proper execution is a condition
for exact integration of the prestressing cable with the prestressed
constuction and the protection of cables against possible corrosion.
In order to meet these requirements properly injection grouts should
posses an approriate strength and resistance to frost.

The basic condition of success of that operation is an accurate
selection of, from the point of so called 'injectability',
components of the cement suspension. Until now the selection has
been done experimentally. If, however, the injectability is defined
as a complex feature determining the ability of cement paste to
fill the cable channel comopletely at its minimum sedimentation then
it will turn out that the optimum estimators of grout properties are
its rheological characteristics.

Great practical advantages emerging from this approach require,
however, an adequately precise recognition and description of
rheological features of cement paste and a knowlege of exact

dependence describing the grout flow in a channel as an effect of its geometry, rheological characteristics of grout and the pressure of the pumping system.

2 Rheological model of injection grout

Investigation made until now have shown that cement grout is characterised by very complicated rheological properties [1] [2] [3] [4] [5]. On the basis of a shearing test in the system of variables shearing stress - shearing rate (τ - $\dot{\gamma}$) for a time point t_i = const the character of its flow curve identifies it as a plastic viscous body diluted by shearing. In general, flow curve τ - $\dot{\gamma}$ shows in this case nonlinearity which depends on concentration of solid phase c_v, cement specific surface F_B, clinker mineralogical composition (contents of C_3S, C_2S, C_3A, C_4AF) and other factors [2] [3] [5]. Because of the lack of rheostability in paste the shearing test has to take into account an additional variable t (time) in relation to which further dependencies show strong nonlinearity.The equilibrium test, which leads to estimation of the general flow function of cement paste,

$$\tau = f(\dot{\gamma}, t) \tag{1}$$

as the two - members dependence

$$\tau = \varphi \ (\dot{\gamma})\cdot\Theta \ (\dot{\gamma} \ t) \tag{2}$$

seems to be the most conclusive and convenient. It enables the clear separation of viscoplastic properties of the paste as a flat component (τ - $\dot{\gamma}$) of the flow function from its tixotropic properties described by the component Θ.

The complexity of the general model (1) causes the necessity of reducing it to a linear one for the sake of practical application. The efect of tixotropic fluidization is seemingly eliminated here by the assumption, in the samples of equilibrium flow, a time point t_i. It may be, for instance, the time point t_1 which is the instance of the debut stress manifestation τ_1 in the beginning of paste shearing. Or it may be the time point t_m corresponding to m-th second of the equilibrium test in which minimum shear τ_m corresponds to the destruction state limit of the paste structure.

The assumption allows the reduction of the complicated real rheological model of paste (2) to Bingham's linearizeed form

$$\tau_i = (\tau_{oi} + \eta_{pi} \cdot \dot{\gamma}_i) \qquad \text{for } i \in (1,\dots,m) \tag{3}$$

The substantial investigation programme comprising 20 cements of different mineral composition and different specific surface was performed. It enabled to specify the dependencies describing the rheological parameters of the equation (3) as a function of material and technology dependent factors. For the tixotropic stage of fluidization corresponding to time point t_m this dependence has the form

$$\tau_o = (1 + \beta) \ F_B \frac{c_v^{(p+1)}}{c_{vg} - c_v} \quad [Pa] \tag{4}$$

$$\eta_p = 7 \ F_B \ (\omega - \omega_g)^{-2.5} \ 10^6 \quad [Pa \ s] \tag{5}$$

$$c_v = V_c \ (V_c + V_w)^1 = (1 + \omega\rho_c)^1 \tag{6}$$

$$c_{Vg} = (1 + \omega_g\, \rho_c)^{-1} \tag{7}$$

$$\omega_g = 0.876\, \omega_o \tag{8}$$

$$p = 3F_B{}^{0.05} M_{sA}{}^{0.03} \tag{9}$$

$$M_{sA} = \frac{C_3S + C_2S}{C_3A + C_4AF} \tag{10}$$

The amount of phase components is in % ;
ρ_c - mass density of cement $[\mathrm{kg/m^3}]$;
F_B - Blaine's specific surface $[\mathrm{m^2/kg}]$;
ω - w/c water - cement ratio;
ω_o - Vicat's specific water;
V_c, V_w - absolute volume of components;
c_{Vg} - limit of concentration of cement volume (acc. to (8)).

3 The flow of the injection grout in a cable channel

The volumetric flow of the rheostable liquid in a pipe of diameter D
in which shearing stress on the wall is τ_w has been described

$$Q = \frac{\Pi\, D^3}{8\, \tau_w} \int_0^{\tau_w} \tau^2 f\,(\tau)d\tau \tag{11}$$

This equation, when Bingham's model for identification of liquid is
introduced, is transformed into very complex Bingham - Reiner's
dependency. If, however, the segment in the fourth power is omitted,
a simple and sufficiently accurate equation is obtained

$$\Delta p = 32\, L/\Phi\ (\eta_p/\Phi \cdot v_a - \tau_o/6) \tag{12}$$

The pressure drop Δp is caused by the resistance to flow of grout
with rheological parameters τ_o, η_p on the length L of cable channel
of reduced diameter ϕ and with average linear rate v_a.
Admitting, therefore, an experimentally determined non-turbulent
flow rate of grout which fills the channel as $v_a < v_1 = 0.2$ m/s,
the pressure drop Dp of grout with rheological parameters as below

$$\{\tau_{oi}, \eta_{pi}\} = \psi\ (w/c,\ \rho_c,\ F_B,\ M_{sA}) \tag{13}$$

and known L and F can be estimated. As a modeled stage of flow a
limiting stage of complete filling of the channel by grout under the
pressure Dp is, obviously, assumed.

4 Optimizing criteria of injection grout composition

The standard failure condition for the hardened grout in a cable
channel has the form

$$R_t \begin{cases} R(7) \geq 20\ \mathrm{MPa} \\ R(28) \geq 30\ \mathrm{MPa} \end{cases} \tag{14}$$

The adherence to this condition means usually the same as the adherence to freeze resistance. On accepting the failure condition in the form of equation

$$Rz = \alpha i \cdot Rc \cdot (c/w - 0.5) = \alpha i \cdot Rc \cdot (1/\omega - 0.5) \geq Rt \qquad (15)$$

where Rc is the cement standard strength, αi is an empirical coefficient and ω is a mass ratio of water and cement, we obtain, from (14) and (15) the following alternative dependence

$$\omega 1 t = \left\{ \begin{array}{c} \omega 11 \\ \omega 12 \end{array} \right. \leq \left(\frac{2\alpha i \ Rc}{2 \ Rz + \alpha i \ Rc} \right)^{-1} \qquad (16)$$

The second condition of limited sedimentation
$$S3h \leq 1.5 \ \%$$
$$S24h \leq 2.0 \ \% \ , \qquad (17)$$
when one of the equation of Wishes - Rendehen, Keizar or Mierzwa [5] is applied gives the double condition as well

$$\omega 2 t = \left\{ \begin{array}{c} \omega 21 \\ \omega 22 \end{array} \right. \leq \omega = \varphi \ (\ldots\ldots S3, S24) \qquad (18)$$

The last condition is the criterion of injectability derived from the equation (12) which leads to a coefficient ω for which rheological parameters (τo, ηo) of the paste enable its complete flow along a cable channel observing the rule

$$\Delta pef < 0.5 \ \Delta pu^{\cdot} \qquad (19)$$

where Δpu is the maximum working pressure which can be achieved by an injecting unit.
On the basis of (3) (4) (5) (6 ÷ 10) (12) (16) (18) the optimum value of the w/c ratio of the injecting grout using the method of succesive approximations and according to the algorithm on fig.1 can be estimated .The grout complies with all the necessary requirements of a proper injection.

5 The influence of cements on injectability

The carried out experiments proved that such seemingly minor factors like specific surface or mineralogical composition have great effect on the process of injection. The effective pressure of the injection of cable channel 30 m long, 32 mm internal diameter Φ and a cable 12 x 7 mm can differ to up to 100 % for different w/c ratio if, for instance, specific surface FB = 300 m^2/kg and the index of mineral composition (10) will be changed within limits MsA = 3.15 to MsA = 5.15 which is shown on fig.2.

6 Conclusion

6.1 Introducing the rheological parameters of injection grouts as a basis of optimising the process of injection creates, as it seems, the new and wider possibilities of prognosing the technical conditions of injection. This procedure can be presented in the form

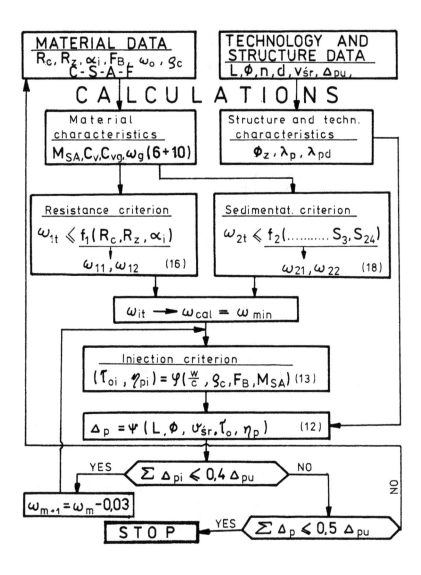

MATERIAL DATA
$R_c, R_z, \alpha_i, F_B, \omega_o, \mathcal{G}_c$
C-S-A-F

TECHNOLOGY AND
STRUCTURE DATA
$L, \phi, n, d, v_{śr}, \Delta_{pu},$

CALCULATIONS

Material
characteristics
$M_{SA}, C_v, C_{vg}, \omega_g (6 \div 10)$

Structure and techn.
characteristics
$\phi_z, \lambda_p, \lambda_{pd}$

Resistance criterion
$\omega_{1t} \leqslant f_1 (R_c, R_z, \alpha_i)$
ω_{11}, ω_{12} (16)

Sedimentat. criterion
$\omega_{2t} \leqslant f_2 (\ldots\ldots S_3, S_{24})$
ω_{21}, ω_{22} (18)

$\omega_{it} \longrightarrow \omega_{cal} = \omega_{min}$

Injection criterion
$(\mathcal{T}_{oi}, \eta_{pi}) = \psi (\frac{w}{c}, \mathcal{G}_c, F_B, M_{SA})$ (13)

$\Delta_p = \psi (L, \phi, v_{śr}, \mathcal{T}_o, \eta_p)$ (12)

YES $\sum \Delta_{pi} \leqslant 0,4 \Delta_{pu}$ NO

$\omega_{m+1} = \omega_m - 0,03$

STOP   YES $\sum \Delta_p \leqslant 0,5 \Delta_{pu}$ NO

Fig.1.Optimum w/c ratio estimation flowchart.

of an algorithm.

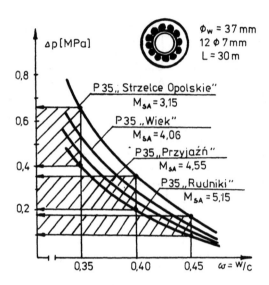

Fig.2. Mineral composition - effective pressure.

6.2 The very important feature of the injection grout is its
liquidity (injectability) and the substantial effect on it has the
mineral composition of cement. The change of the MSA modulus within
the range from 3 to 5, when other factors like Rc and FB are
constant , can result in the change of rheological parameters of
injection grout of several score %. It can have essential
significance for technical results of the process of injection.

7 References

[1] Tattersall G.H. (1976) The workability of concrete.Sheffield COR
[2] Tattersall G.H. and Banfill P.F.G. (1983) The rheology of fresh
    concrete. Pitman Adv. Publ. Program. London
[3] Berg vom W. (1979) Influence of specific surface and
    concentration of solids upon flow behaviour of cement pastes. Mag.
    Con. Res. vol.31 no.109
[4] Mierzwa J. (1986) Tixotropic - plastic - viscous rheological
    model of cement paste as the result of physical and mineralogical
    cement properties. 8-th Congress Chem. Cem. Rio de Janeyro (274 ÷
    278)p.Ⅵ.
[5] Mierzwa J. (1988) Własności reologiczne zaczynow cementowych
    stosowanych w wybranych procesach technologicznych budownictwa.
    Monografia 69. Politechnika Krakowska. 1988.

22 INVESTIGATION OF THE RHEOLOGICAL PROPERTIES AND GROUTABILITY OF FRESH CEMENT PASTES

W. ZIMING, H. DANENG and X. YAOSHENG
China Building Materials Academy, Beijing, China

Abstract
Improvement of the groutability of cement materials is the krux to widen the uses and applications of cement grouting technique. This paper introduces the systematic study of the stability and rheological properties of cement paste, and discusses the factors which influence the mentioned properties of fresh cement pastes as well. Analyses of the direct relation between rheological properties and groutability about cement pastes were made according to rheological theories. The formula of groutability of cement paste was proposed to describe the relation of groutability with sedimentary velocity and plastic viscosity of cement paste quantitatively. The effective measures to improve the groutability of cement paste are to increase the fineness of cement, to use cement pastes with optimal water cement ratio and to add suitable superplasticizers.
Keywords: Grouts, Groutability, Stability, Rheology, Cement Paste, Sedimentary Velocity, Superplasticizer.

1 Introduction

Since French engineer Clarles Berigry (1772-1842) first used grouting method in the rehabilitiation of Dieppe Dam, Cement grouting technique has been widely used in the fields of construction, metallurgy, mines, dam etc. cement has some advantages compared with organic grouting materials regarding to its cost, resourse, and its properties. But the poor grountability of cement paste is the most important reason that obstructs the replacement of polymer grouting materials. Attention has been paid to improve the groutability of cement paste and developments have been achieved in some areas. Following problems must be solved before any remarkable progress is realizable.

The definition of groutability of cement paste and its implication. Usually grouting ratio C (D85/d15) was used to evaluate the grouting effect of cement paste. To acquire satisfactory grouting results, C should be in the range of

6 to 25. C varies considerablely with circumstance condition.

Rheological properties of high W/C ratio cement paste. The properties of normal W/C ratio cement pastes have been studied thoroughly. But the rheological properties of cement paste with high W/C ratio has not been adquately studied.

The relationship of stability and groutability of cement paste. Grouting practice has proven the importance of the stability of cement paste in grouting engineering, clombandi pointed out that uniform cement pastes make cracks in rocks filled completely and render a strong bondness between hardened cement paste and rocks, therefore improve the durability and decrease the permeability of project. X.Y. Chen proposed that groutability is the ability that cement paste penetrates into cracks of rock and sets normally inside cracks. This paper introduces the systematic studies on the rheological properties, stability and groutability of cement pastes.

2 Materials and method

2.1 Materials and apparauts
Portland cement was prepared by grinding the mixture of 95% cement clinker and 5% gypsum into required finenes. The clinker composition was shown in Table 1.

Table 1. Composition of clinker.

Composition	CaO	SiO_2	Al_2O_3	Fe_2O_3	MgO	SO_3	K_2O	Na_2O	IOL
Content	67.0	19.48	6.08	4.09	0.92	0.05	0.57	0.16	1.54

Superplasticizer UNF-II and LGS were used in the experiments. Rotating coaxial cylinder viscomete NXS-11 was used in measuring rheological parameters of cement pastes with W/C ratio below 1.0. As to the cement pastes with W/C ratio above 1.0, rheological parameters were determined by Vertical capilliary viscometer (Fig.1). Rheological parameters are calculated from Buckingham & Reiner equation.

$$\tau = \eta_{pl}\frac{4Q}{\pi R^3} + \frac{4}{3}\Theta t$$

Laser particle distribution detector was taken to analyse the particle distribution of cement paste.

2.2 Measurement of sedimentary velocity
The stability of suspension is defined as the ability to maintain uniform from the point of view of physicochemistry. The stability of cement paste can be represented by two

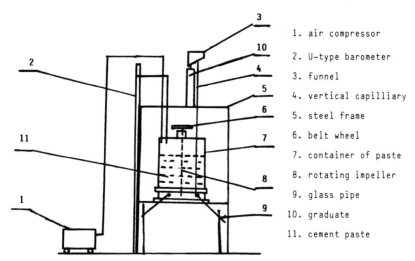

1. air compressor
2. U-type barometer
3. funnel
4. vertical capilliary
5. steel frame
6. belt wheel
7. container of paste
8. rotating impeller
9. glass pipe
10. graduate
11. cement paste

Fig.1. Diagram of vertical capillary viscometer.

factors-bleeding and particle separation, which is affected by W/C ratio of cement paste, the shape and size of cement particle. Dynamic stability is the ability to keep uniform when cement paste is in flowing condition. According to Thomas, sedimentary velocity V_C is the speed at which a particle layer is formed at the bottom of a pipe when cement paste flows through the pipe. Apparently V_C represents the stability of flowing paste. V_C was determined by observing the paste flowing through a transparent glass pipe at controlled speed.

3 Results and discussions

3.1 Rheological properties of grouting cement paste
The influences of cement fineness and W/C ratio of paste on rheological parameters are shown in Table 2.

Table 2. Influence of fineness and W/C on rheological parameters.

W/C	0.5		0.8		1.0		2.0		3.0		5.0	
S (cm^2/g)	η_{pl}	T_o	η_{pl}	T_o	η_{pl}	T_o	η_{pl}	T_o	η_{pl}	T_o	η_{pl}	T_o
3475	36.0	66.0	9.6	8.8	4.7	5.6	2.6	2.1	1.8	2.2	1.5	1.3
4487	67.0	128.0	12.5	20.0	11.6	9.4	3.8	2.6	2.5	1.2	1.9	0.7
5213	78.0	136.0	49.0	49.6	15.6	13.5	3.8	4.4	3.6	3.4	2.2	1.9

The characteristics that n_{pl} and T_0 vary with the W/C ratio can be represented with following equations.

When $S=5213 \text{ cm}^2/\text{g}$

$$n_{pl} = 1.104 \text{EXP}[11.70/(1+3.13\text{W/C})] \quad (R=0.97) \quad (1)$$
$$T_O = 0.830 \text{EXP}[13.16/(1+3.13\text{W/C}] \quad (R=0.99) \quad (2)$$

n_{pl} and T_0 change with W/C ratio in exponent law. When W/C ratio is above 2.0, specific surface of cement shows little affection on rheological parameters.

Effect of superplasticizer on the rheological parameters are shown in Table 3.

Table 3. Effect of superplasticizer on T_0 and n_{pl}.

W/C	0.5		0.8		1.0		2.0		3.0		5.0	
Type	n_{pl}	T_0	n_{pl}	T_0	n_{pl}	T_0	n_{pl}	T_0	n_{pl}	T_0	n_{pl}	T_0
Reference	78.0	13.6	49.0	14.9	14.9	15.2	3.8	4.4	3.7	3.4	2.2	1.9
LGS	24.0	4.8	17.0	10.0	1.6	4.1	2.3	1.6	1.9	0	1.4	1.0
UNF-II	24.5	–	17.0	11.2	11.2	3.0	2.7	1.2	2.4	0.4	1.9	0.3

Clearly, superplasticizer improve the rheological properties of cement paste to certain degree. But again when
W/C exceed 2.0, admixtures show little effect. Yielding
value are roughly constant after adding superplasticizer
in experimental condition. The lower the W/C, the greater
the effect of admixture on rheological parameters.

Practical grouting process often lasts up to 4 hours and
in the process cement paste is contantly rotated. Rheological parameter changing with mixing time was tested to
evaluate the grouting practice. Results are shown in Table
4.

Table 4. Change of rheological parameters with mixing time.

W/C	0.8		1.0		3.0		5.0	
t(min.)	$\dfrac{n_{pl}}{n_{pl}'}$	$\dfrac{T_0}{T_0'}$	$\dfrac{n_{pl}}{n_{pl}'}$	$\dfrac{T_0}{T_0'}$	$\dfrac{n_{pl}}{n_{pl}'}$	$\dfrac{T_0}{T_0'}$	$\dfrac{n_{pl}}{n_{pl}'}$	$\dfrac{T_0}{T_0'}$
30	100	100	100	100	100	–	100	100
60	104	102	133	100	–	–	–	–
120	314	162	162	160	–	–	–	–
180	510	227	183	207	110	–	103	80
300	–	–	–	–	134	–	118	51

T_0' and n_{pl}' represent the plastic viscosity and yielding value at 30 min. When W/C is below 1.0, n_{pl} and T_0 increase sharply with the mixing time. So practical grouting process should not last too long when low cement paste is used.

.2 Sedimentary velocity V_C

edimentary velocity is also affected by W/C of paste, ement specific surface and type of admixture. Test esults are shown in Table 5, and 6.

$V_C=1$ means that paste bleeds only in still state. From he test results, we can reach the conclusion that superlasticizers decrease the stability of cement paste in ense of bleeding and particle segration.

'able 5. V_C of various cement paste.

V_C (cm/s) W/C S (cm^2/g)	0.5	0.8	1.0	2.0	3.0	5.0
3470	1	1	11	50	55	59
4481	1	1	1	10	26	37
5213	1	1	1	8	20	31

able 6. Affection of admixture on V_C.

V_C (cm/s) W/C Type	0.5	0.8	1.0	2.0	3.0	5.0
reference	1	1	1	8	20	31
UNF-II	1	1	1	23	31	36
LGS	1	1	1	18	31	44

Evaluation of groutability

It has long been discussed that what the implication of groutability is. The author hold that implication of groutability about cement paste is the comprehensive effects of rheological properties and stability. Groutability is a intrinsic property of cement paste in the same way of rheological property.

4.1 Formula of groutability
Groutability (G) can be expressed with following formula.

$$G = K \frac{1}{n_{pl} V_C} \qquad (3)$$

K is a constant which is determined by width of crack (a)
and grouting material pardicle size (d15)

$$K = 0 \qquad \text{when } a/d15 < 5$$

$$K = 1 \qquad \text{when } a/d \geqslant 5$$

when $a/d15 < 5$, paste can not be grouted into crack   duing
particle filtering effect.

Unit of G is cm/dyne. it means the flowing distance
under per unit force in uniform condition.

Applying formula (3) into test results. The groutability of cement paste is shown in Fig.2, Fig.3.

Groutability reaches its acme when W/C=1.0. Superplasticizer improve the groutability of cement pastes when its
W/C ratio is below 1.0.

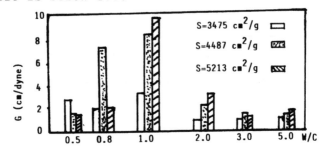

Fig.2.   Relationship of groutability with
          W/C and specific surface.

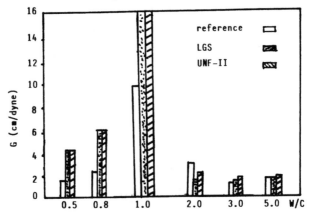

Fig.3. Effect of superplasticizer on groutability.

5 Conclusions

Groutability of cement paste upgrades quickly with the
increase of W/C ratio in range of 0.5 to 1.0. When W/C is
above 2.0, groutability decreases with the increase of W/C

ratio. Fine cement pastes have better groutability than normal cement pastes. Superplasticizer improves the groutability of cement paste when W/C is below 2.0. The optimal W/C ratio is between 0.8-2.0. The effective measures to improve the groutability of cement paste are to increase the fineness of cement, to use cement paste with optimal W/C ratio and to add suitable superplasticizer.

Reference

R. Glossp, "Opending Address"; Symposium on Grouts and Drilling Muds in Engineering Practice, London.
Huang Daneng, "Rheology of Cement and Concrete", Beijing (1981)
Ma Zhigui, "Grouting Characteristics of Sand Base"
A.H. Adamovich, "Grouts in Marine Construction" (1985)
X.D. Yang, Groutability of Cement Paste, Master Thesis (1985)
X.R. Chen, "Some Problems about Grout in Rock base with Fine Crack"; Procedding of Marine Construction Practice.
A. Mayer,"Modern Grouting Technique"; Symposium on Grouts and Drilling Muds in Engineering Practice, London (1963)
Wasp E.J., Pipe transportation of solid materials, (1977)

PART FIVE
MORTARS

3 A COAXIAL CYLINDERS VISCOMETER FOR MORTAR: DESIGN AND EXPERIMENTAL VALIDATION

P.F.G. BANFILL
University of Liverpool, School of Architecture and
Building Engineering, Liverpool, UK

Abstract
The design requirements - sample size, gap size, cylinder radius ratio
and surface profiling - of a coaxial cylinders viscometer for mortar
are considered and an apparatus which meets them is described.
Calibration experiments with model fluids show that it performs
satisfactorily. Mortar conforms to the Bingham model and has yield
values and plastic viscosities intermediate between those of cement
paste and fresh concrete. Mortar also undergoes structural breakdown
under shear, which has implications for mixing and testing methods.
Keywords: Mortar, Viscometer, Theory, Calibration.

1 Introduction

We may define mortar as a fine grained mixture of binder, sand and
water with a maximum particle size of about 2mm. The binder may be
cement, lime or gypsum plaster with other mineral, polymeric or
chemical admixtures. Thus it includes bricklaying mortars; polymer
modified mortars for repair, patching or filling, whether premixed and
packaged or mixed at the point of use; floor screeds and levelling
compounds; and decorative or waterproofing finishes.
    The rheology of such materials is important for several reasons.
Quality control of production can be carried out at the time of use
rather than waiting for results of tests on the hardened material: a
correct definition of the rheology of the materials enables simple,
soundly based and informative control tests to be devised. A
rheological consideration of the conditions of use may also provide
information useful to those involved in the formulation of materials.
In practice the rheology of the mix must be right for its required
purpose otherwise the workman will do a poor job. If it is not right
he will probably change the product by adding more or less water with
consequent changes in performance. Finally, the rheology of mortar
may provide information relevant to understanding the rheology of
fresh concrete.
    A very large number of empirical tests have been used for mortars,
but, as for fresh concrete, very few have been widely used and even
these are accepted as having limitations. In view of the success of
the two-point approach to the rheology of fresh concrete (Tattersall

and Banfill, 1983), which has shown that it is a Bingham plastic needing to be characterised by two parameters, the yield value τ_o and the plastic viscosity μ

$$\tau = \tau_o + \mu \, \dot{\gamma} \qquad (1)$$

τ = shear stress, $\qquad\qquad \dot{\gamma}$ = shear rate

it seemed logical to use the concrete apparatus for mortar. In early trials the two-point apparatus proved to be insufficiently sensitive and even modifications to the impeller were unsuccessful. The relatively small grain size of mortars suggested that a coaxial cylinders viscometer could be used and this paper describes such a viscometer developed from the two-point workability apparatus.

2 The coaxial cylinders viscometer

2.1 Governing equations

In a coaxial cylinders viscometer the material under investigation fills the annulus and also the disc-shaped space under the inner cylinder. When one cylinder is rotated at angular velocity Ω the viscous drag exerted by the sample tends to rotate the stationary cylinder and the torque, T, is measured as that required either to rotate the moving cylinder or to hold the other cylinder stationary. The torque is the sum of that due to shear in the annulus, T_a, and in the disc under the inner cylinder, T_b. The flow in the annulus for a sample conforming to the Bingham model is described by the Reiner-Rivlin equation:

$$\Omega = \frac{T_a}{4\pi\mu h_a} \left(\frac{1}{R_b^2} - \frac{1}{R_c^2} \right) - \frac{\tau_o}{\mu} \ln\left(\frac{R_c}{R_b} \right) \qquad (2)$$

(h_a height of cylindrical annulus, R_b and R_c radius of inner and outer cylinders, respectively). The flow under the flat bottom of the inner cylinder is described by the equation for a parallel plate viscometer:

$$T_b = \frac{2}{3}\pi \tau_o R_b^3 + \frac{1}{2}\frac{\pi \mu \Omega}{h_b} R_b^4 \qquad (3)$$

(h_b height of the gap under the inner cylinder). Combining equations (2) and (3) to give the total torque T produces

$$T = \pi \tau_o\left[\frac{4 h_a}{(1/R_b^2 - 1/R_c^2)}\ln\frac{R_c}{R_b} + \frac{2}{3}R_b^3\right] + \pi\mu\Omega\left[\frac{4 h_a}{(1/R_b^2 - 1/R_c^2)} + \frac{R_b^4}{2 h_b}\right] \qquad (4)$$

Equation (4) is of the form $T = A + B\Omega$ where A and B are proportional to τ_o and μ respectively and the constants of proportionality depend only on the geometry of the system. It can therefore be used to determine τ_o and μ from the straight portion of a graph of T against Ω .

2.2 Geometrical requirements

For particulate materials, such as mortar, various well-established rules exist concerning the dimensions of a viscometer which will give measurements to a satisfactory degree of accuracy. The gap size should be not less than ten times the diameter of the largest particles in order to minimise the effect of changes in particle packing near the walls. The radius ratio should be as near to 1.0 as possible and not greater than 1.2 to ensure a small variation in shear rate across the gap and to minimise the speed range at which plug flow occurs. The height/radius ratio should be not less than 1.0 to minimise the contribution of the bottom of the inner cylinder. Finally, to minimise the risk of slippage at the cylinder walls it is customary to profile them in some way, in which case the asperities in the profile should be no smaller than the largest particle in the material.

The combination of minimum gap width, radius ratio and height/radius ratio means that the smallest practicable viscometer for a mortar with particles of 2mm maximum size has R_b = 100mm, R_c = 120mm and h_a = 100mm, but in addition the sample size must be sufficiently large to minimise sampling errors and unrepresentativeness of the sample. Also a sufficiently large portion of the total sample taken must be subjected to the conditions of the test to minimise errors due to segregation and bleeding between the tested part and any "dead" part. Such a dead part may arise if segregation results in a slippage layer under the inner cylinder. It is suggested that target values are a total sample volume of 5 x 10^{-3} m^3 with not more than 25% of the sample being in dead space, not taking part in the test.

The volume of the annulus in the smallest practicable viscometer mentioned above is given by:

$$V_a = \pi h_a \left(R_c^2 - R_b^2 \right) \qquad (5)$$

and for h_a = 100mm, R_b = 100mm and R_c = 120mm, V_a = 1.38 x 10^{-3} m^3. Therefore any value of h_a from 100 to 350mm will be within the range of reasonable sample sizes. Similarly, the volume of the bottom space in the smallest practicable viscometer is:

$$V_b = \pi h_b R_c^2 \qquad (6)$$

For h_b = 20mm, V_b = 0.9 x 10^{-3} m^3. If this is taken to be 25% of the total sample volume then V_a = 2.7 x 10^{-3} m^3 and h_a must be at least 200mm. However, h_a more than 250mm would require major alterations to the frame of the two-point apparatus.

In conclusion the geometrical and volume requirements are met by R_b = 100mm, R_c = 120mm, h_b = 20mm, h_a = 200mm.

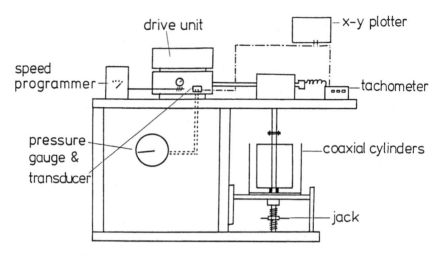

Fig. 1. General arrangement of apparatus.

3   Description of apparatus

The coaxial cylinders viscometer, as built, is shown schematically in Fig. 1. It consists of an electric motor driving the inner cylinder through a variable speed hydraulic gear (Carter Gear F10, Renold Ltd) and a 5.2:1 right angled gear box. The inner cylinder is mounted coaxially in the outer cylinder, with bearings below to ensure accurate alignment. The speed of rotation is measured automatically by an electronic tachometer (Compact Instruments) and the torque is proportional to the hydraulic pressure developed in the Carter Gear so that a simple 0-1000 lb/in^2 (0-7 N/mm^2) pressure gauge gives a direct visual reading of torque. In addition, a pressure transducer (0-1000 lb/in^2, RDP Electronics), mounted in the pipe connecting the Carter Gear and gauge, gives an electrical signal which can be combined with the signal from the tachometer to give a direct X-Y plot on a potentiometric recorder.

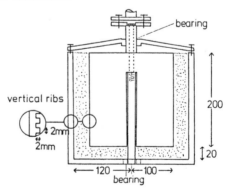

Fig. 2. Detail of coaxial cylinders.

Table 1. Dimensions of the viscometer.

Parameter	Recommended Value	Value in this apparatus
$(R_c - R_b)$/max particle size*	10	10
R_c/R_b	1.2	1.2
h_a/R_b	1	2.0
ribs/max particle size*	1	1.0

* assumed to be 2mm

The detail of the cylinders is shown in Fig. 2. Important features are the vertical ribs machined in the cylindrical surfaces and the radial ribs machined in the flat bottom surfaces of both cylinders. Table 1 is a comparison of the actual dimensions with the ideals described in Section 2.2.

Finally, in order to be able to produce conveniently a flow curve of torque against speed, a controller programmes the change of speed automatically. The speed control on the Carter Gear is a handwheel which rotates a spindle which moves a valve within the gear. 4.5 revolutions are needed to change the output rotation from stationary to its maximum speed of 15.6 rev/sec (3 rev/sec of inner cylinder), therefore an auxiliary electric motor is used to drive this spindle at a constant speed to give the required acceleration or deceleration of the inner cylinder. Limit switches prevent overspeeding. This speed control was set up so that a complete speed cycle (from stationary to maximum and back) could be achieved in 1, 2, 5, 10 and 20 minutes. This arrangement permits a flow curve, expressed as raw data of torque and speed, to be produced automatically.

4   Calibration of the apparatus

This calibration was done in two parts: firstly the speed and torque readings were checked, and secondly the flow curve of a known fluid was determined.

The speed calibration was simply confirmed visually with a stopwatch. The torque-pressure relationship was determined by mounting a lever arm on the vertical shaft where the inner cylinder is normally mounted. The lever arm was fitted with a plummer block which could be tightened on the shaft and the force on the lever arm was indicated by a spring balance anchored to an immovable object. This arrangement gives the torque in the form of a braking force, which is linearly related to the pressure in the hydraulic gearbox as shown in Fig. 3. From this the torque (Nm) is given as 0.0272 times the

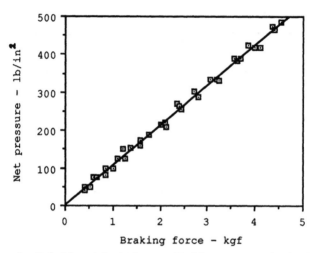

Fig. 3. Relationship between fluid pressure in hydraulic gearbox
and torque.

pressure in lb/in^2 .

The torque-pressure relationship depends on the temperature of the
fluid inside the gearbox, so it is always necessary to run the
apparatus until it is warm before attempting to obtain a flow curve.
Additionally, there are various frictional losses in bearings and
gears which cause pressures to be indicated when the apparatus is
running with no sample in the gap. Trials showed that the idling
pressures became constant after 30 minutes and conformed to the
relationship:

$$P \text{ (idling)} \quad = \quad 170 \quad + \quad 4.12 \, \Omega \tag{7}$$

This idling pressure line must be subtracted from the experimentally
obtained total pressure-speed relationship in order to give the net
pressure and hence the torque-speed relationship due to the viscous
drag of the sample in the gap.

A Newtonian fluid (high viscosity oil) was used to check that these
calibration constants for the apparatus gave the correct value of
viscosity. From the dimensions of the cylinders using equation 4 the
viscosity, η , of such a fluid is given by the slope of the graph of
T against Ω

$$\eta = \quad 11.80 \quad x \quad \frac{T}{\Omega} \tag{8}$$

Results for this oil were obtained at three temperatures and the
viscosity was also determined using a Haake Rotovisco RV2 coaxial
cylinders viscometer. Table 2 summarises the data, which agree within
10%. This overestimate of viscosity is considered to be acceptable in
view of the following possible sources of error in deriving the
results. The oil pressure fluctuates with a short time period,
providing a source of error in assessing the slope of T/Ω .
While great care is taken in ensuring that the hydraulic drive unit

Table 2. Calibration checks with a Newtonian oil.

Temperature oC	Slope T/Ω	Viscosity (Ns/m^2) (eqn. 8)	(Rotovisco)
27.5	1.074	12.67	11.67
28.1	1.019	12.02	10.84
28.5	0.981	11.58	10.38

.s fully warmed up before starting the test, if the idling relationship lies to higher torques than that given in equation 7 the viscosity will be overestimated. Finally equation 7 was obtained with the inner cylinder removed: additional friction in its bearings will bias the readings towards higher torques, again overestimating the viscosity.

Specimen results

1:3 cement:sand mortars were tested using a range of total cycle times from 1 to 20 minutes. The raw results were processed as shown schematically in Fig. 4, and the yield values and plastic viscosities calculated according to equations 8 (above) and 9:

$$\tau_o = \quad 58.5 \quad x \quad T_o \tag{9}$$

where T_o is the extrapolated intercept on the torque axis. The effect of cycle time is shown in Fig. 5.

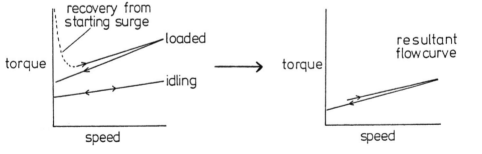

Fig. 4. Stages in deriving a flow curve.

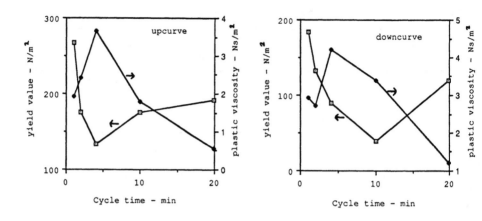

Fig. 5. Effect of cycle time on yield value and plastic viscosity.

1:4, 1:5 and 1:6 cement:sand mortars containing various admixtures were prepared at a standard consistence (10.0 ± 0.5mm ball penetration according to B.S. 4551) by adjusting the water content. They gave the results shown in Table 3.

Table 3. Rheological parameters of various mortars (downcurves only)

Mix	Cement: Sand	Water/ Cement	Ball pen mm	Yield value N/m^2	Plastic viscosity Ns/m^2	Notes
AE1	1:4	1.10	10.5	67	2.7	Air entraining
AE2	1:5	1.20	10.0	147	1.9	mortar
AE3	1:6	1.35	10.0	103	1.3	plasticiser (3 ml/kg) cement)
WP1	1:4	1.05	10.0	100	2.8	Stearate- based water
WP2	1:5	1.30	10.0	100	2.8	-proofer (5 ml/kg
WP3	1:6	1.45	10.0	120	3.8	cement)
PM1	1:4	0.75	10.0	190	1.6	Polymer latex dis-
PM2	1:5	0.85	10.0	40	1.6	persion (200 ml/kg
PM3	1:6	0.95	10.0	30	1.2	cement)

6   Discussion

The results reported here confirm that fresh mortar conforms to the Bingham model and needs two parameters to define its rheological properties, the yield value and plastic viscosity. Additionally mortar shows structural breakdown and in every test reported here the downcurve was on the lower torque side of the upcurve. This breakdown is shown particularly clearly by the range of yield values determined. The upcurves gave between 100 and 310 N/m^2 while the downcurves gave 30 - 180 N/m^2. This structural breakdown is caused by shear during the course of the test and suggests that the yield value and plastic viscosity of mortar determined in the apparatus will be affected not only by the test itself but also by the mixing technique used in preparation and by any shear during handling prior to testing. Fig. 5 confirms that the progressively increasing work of shearing done on the mortar as the cycle time increases to 4 and 10 minutes reduces the yield value of both up and downcurves. By analogy with the known effect of cycle time on the flow curve of cement paste, which also breaks down (Banfill & Saunders, 1981), it is expected that short cycle times will produce more reproducible results. Short cycle times are also more convenient and are not prone to the errors due to segregation of the sample during the test.

The results of the limited number of tests on mortars show that the influence of composition is complex. By analogy with fresh concrete this is not unexpected. However, there is also clearly no correlation between the empirical dropping ball test result and the yield value or plastic viscosity. The yield values and plastic viscosities of mortar fall between those reported in the literature for cement paste and fresh concrete. This seems reasonable in view of the likely contribution of the aggregate in the material: as aggregate size increases a greater proportion of the externally applied stresses can be borne by the aggregate and so the material becomes stiffer. It also confirms ordinary subjective observation of the fresh materials.

The apparatus in its present form has various shortcomings which will limit its widespread application. It is rather difficult to load the sample into the viscometer and time consuming to clean it after a test, because of the bearing underneath the inner cylinder. This bearing is needed to ensure that the inner cylinder rotates coaxially within the outer cylinder. Any eccentricity makes the gap width fluctuate as the cylinder rotates, pumping mortar up and down which loses contact with the inner cylinder. It is also impossible to obtain a full flow curve including low shear rates on the upcurve because the hydraulic gearbox gives a surge of pressure as it starts. This means that indicated torques are unreliable at speeds below 0.5 rev/sec.

Nevertheless, this is the first successful and correctly dimensioned coaxial cylinders viscometer for a cement-based material containing aggregate particles. The good agreement with the viscosity of a known fluid also means that a high level of confidence can be attached to the results for mortar. Having established that the Bingham model applies to mortar it will now be possible to obtain results in any appropriate mixer-type apparatus, which is capable of measuring speed and torque simultaneously, convert them to yield value and plastic viscosity in fundamental units by the theory of Tattersall and Bloomer (1979), and confirm them with this viscometer. In short, the coaxial cylinders viscometer may be used as a reference instrument for checking other more convenient apparatus.

7   Conclusions

The coaxial cylinders viscometer for mortar met the design requirements for a particulate suspension and gave correct results in calibration experiments with a Newtonian fluid. It may be used as a reference when checking other test instruments.

The yield values and plastic viscosities of a range of fresh mortars were between 30 and 310 N/m^2 and 1.2 and 4.0 Ns/m^2 respectively – consistent with those reported for fresh concrete (higher) and cement paste (lower). All flow curves exhibited structural breakdown and short test cycle times are recommended.

8   References

Banfill, P.F.G. and Saunders, D.C. (1981). On the viscometric examination of cement pastes, **Cem. Concr. Res**, 11, 363-370.
Tattersall, G.H. and Banfill, P.F.G. (1983). **Rheology of Fresh Concrete**, Pitman, London.
Tattersall, G.H. and Bloomer, S.J. (1979). Further development of the two point test for workability and extension of its range, **Mag. Concr. Res.**, 31 (109), 202-210.

9   Acknowledgements

I am grateful to B. Atkinson for experimental assistance and to S. Robinson and W. Baker who built the apparatus.

24 THE USE OF THE BRABENDER VISCOCORDER TO STUDY THE CONSISTENCY OF FRESH MORTAR BY TWO-POINT TESTS

F. HORNUNG
E. Schwenk Zementwerke KG, Ulm,
Federal Republic of Germany

Abstract
The Brabender ViscoCorder is a small variable speed viscosimeter designed for use with mortar. The use of a push-button speed-tuner allowed reproducible adjustment to different speeds during each test
which permitted the determination of the flow curves of fresh mortars.
In two-point tests, the influence of different cements, sands and admixtures on the yield value and the plastic viscosity of mortar were
studied.
Keywords: Rheology, Consistency, Workability, Fresh Concrete, Mortar,
Viscosimeter, ViscoCorder.

1 Introduction

The constant workability of fresh concrete for a particular placement
technique is an important requirement for the technical and economical
success of professional concrete production sites. Even though the
concrete may have the specified slump, difficulties can arise on the
site if the consistency of the fresh concrete differs from the expec-
tations of the construction team or does not meet the requirements of
the placement method or the concrete plant provided. At present it is
not possible to measure accurately and reproducibly the workability of
fresh concrete. Thus a large number of experiments are necessary to
produce concretes which comprise different basic ingredients and show
nevertheless comparable workability. Owing to the difficulty in studying the consistency of fresh concrete in a rheometer, one tries to
draw conclusions on the properties of concrete on the basis of
measurements with mortar.
    The Brabender ViscoCorder is commonly used in Germany. It is a
small coaxial cylinder viscosimeter especially designed for use with
mortar. In general, the rotational speed is held constant (Teubert
(1981), Wolter (1985)). Since the consistency of fresh concrete and
mortar is, to a good approximation, described by the Bingham model
(Banfill (1987), Tattersall (1983), (1987)) the so-called two-point
test can be used to study their rheology.

2 Experimental Details

2.1 Description of the ViscoCorder
The ViscoCorder (Fig. 1) consists of a cylindrical container, for the mortar specimen, which is mounted on a rotating turntable. A concentric paddle is mounted on a torque gauge comprising a calibrated spring fixed to a chart recorder pen. As the cylinder rotates, the force due to the friction of the mortar flowing through the blades of the paddle presses against the spring and moves the pen across the chart recorder paper. Full scale deflection (1000 chart units) corresponds to a torque of 1000 cmg (100 Nmm). In the original version of this apparatus, the rotational speed is continuously variable up to 250 rev/min. The speed is adjusted with a calibrated knob and measured with a tachometer.

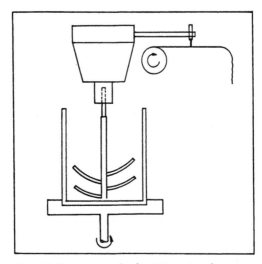

Fig. 1 Diagram of the ViscoCorder

The author exchanged the calibrated knob for a push-button system. This enabled the preselection of different speeds which could be reproducibly selected during the tests. In addition, the rate and sense of the speed change remain the same.

2.2 Materials
In the course of the research programme a wide range of cements, sands and admixtures was tested. Table 1 shows some of the materials used in the tests whose results are mentioned in this report.

Table 1. Cements.

No.	Type		works	spec. surface	
C1:	Pz	35F	Schwenk Karlstadt	3200	cm²/g
C2:	Pz	35F	Schwenk Mergelstetten	3200	cm²/g
C3:	Pkz	35F	Schwenk Mergelstetten	4200	cm²/g

Grading of the cements:

	1	2	4	8	16	32	64	128	192	$*10^{-6}$m
C1:	5.0	11.8	19.3	30.5	47.6	71.9	90.7	98.7	100.0	(%)
C2:	5.6	11.0	20.2	32.0	48.0	69.3	88.2	98.9	100.0	(%)
C3:	7.9	15.2	25.8	38.4	55.1	77.5	94.6	99.8	100.0	(%)

Table 2. Sands.

				size of sieves			
No. type	0.063	0.09	0.125	0.25	0.50	1.0	(mm)
S1: quartz	1.7	2.5	11	41	73	100	(%)
S2: Viscosand*⁾	0.0	0.5	9	36	54	100	(%)
S3: quartz	0.0	0.0	6	13	38	99	(%)
S4: Munich	1.8	-	14	37	70	100	(%)
	total percentage passing through sieve						

*⁾Viscosand is a sand especially produced for use in the ViscoCorder.
Its grading is comparable to the "Normensand DIN 1164", but without
the 1/2 mm fraction. It is packed in bags containing 1000 g.

2.3 Mixing of the Mortar

The DIN 1164 mortar mixer was used. Its blade rotates in planetary
motion at two speeds. The water was weighed in the empty dry mixer
bowl and the cement was then added and mixed for 30 seconds at
140 rev/min. The sand was added 30 seconds later with the mixer kept
running. After 60 seconds from the start of mixing, the speed was
switched to 285 rev/min. The mixing was stopped after a total mixing-
time of 2 minutes.

2.4 Details of the Mortar Mixes

The mix proportions of all mortars were based on 1000 g sand. Cement
and water were calculated according to the required water-cement (w/c)
ratio and the filling ratio (f_v), i.e. the volumetric proportion of
sand to cement paste.

$$\text{filling ratio:} \quad f_v = V_{sand} / (V_{cement} + V_{water}) \tag{1}$$

2.5 Test Procedure

The ViscoCorder was filled with mortar up to the level marker inside
the cylinder. The weight of the sample was noted if water or plastici-

sers were to be added during the test. After 3 minutes from the start
of mixing the ViscoCorder was switched on at 120 rev/min. The speed
was then changed in steps as shown in Fig. 2.

The influence of a higher w/c-ratio on the consistency curve was
investigated by adding water, in relation to the sample weight, at the
end of a cycle of speeds. The next speed cycle was started when constant torque had been reached at 120 rev/min.

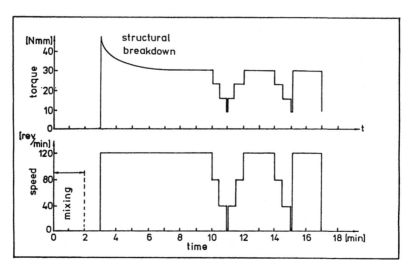

Fig. 2 Construction of the flow curve.

Mortar shows structural breakdown in the period immediately after
starting the ViscoCorder, i.e. the torque decreases when the speed of
rotation is kept constant. Thus to obtain reproducible data, the consistency curve measurement should not be started before the structural
breakdown at the highest shear rate in the test has completed. The
time needed for this depends on the behaviour of the mortar and the
mixing energy brought into it. Owing to the structural breakdown, the
consistency curve for decreasing rotational speed is expected to be at
lower torques than the curve for increasing speed. This effect could
not be resolved with the present set-up. The torques for decreasing
and increasing speed lay within +- 1 Nmm of the average value, but
showed no significant connection with it.

The left-over mortar in the mixer was occasionally tested with the
Haegermann flow table (DIN 1060). This test was performed between 4
and 10 minutes after the start of mixing, during which time the Visco-
Corder had been rotating constantly at 120 rev/min.

Some tests were done to ascertain the transferability of the results obtained by using the ViscoCorder to concrete. Therefore the
slump (ISO 4109) and flow diameter (DIN 1048) of comparable concrete
samples were measured. The concrete mixtures had the same cement, sand
and w/c-ratio as the mortars tested in the ViscoCorder. The filling-
ratios (f_v) of the concretes were 0.80 to 1.07, while those of the
mortars were 0.60 to 0.80.

2.6 Reproducibility and Scatter of the Test Values

The consistency range of mortar tested in the ViscoCorder should produce a torque between 10 and 40 Nmm at 120 rev/min. In the case of mortars with a higher flow resistance unsteady flow conditions caused the recorder pen to vibrate excessively. Mortars of low consistency tend to separation and sedimentation of the coarser sand particles.

A total of 8 mixtures of identical composition were each tested 6 times. The cement used was always from the same sample. A complete bag of the Viscosand was used for the tests. The differences in weight (0 to -3g) were adjusted to 1000g. It was found that during any measurement with several speed cycles, the scatter in results was approximately +- 1 N/mm. The scatter was +- 1.5 Nmm for different tests with the same number of speed cycles. This level of accuracy necessitates a skilled laboratory assistant.

When the speed cycle was repeated several times, the deviation from a straight line for speeds between 40 and 120 rev/min was found to be within the area of experimental error.

2.7 Analysis Method

As already described in 2.1, depending on the speed of the ViscoCorder and the viscous resistance of the mortar, the torque registered is caused by the mortar flowing through the blades of the paddle. The consistency curves were recorded by measuring the torque at 4 different speeds (120, 80, 40, 0 rev/min). By stopping the ViscoCorder during the test it was possible to measure the residual torque 'g_m'.

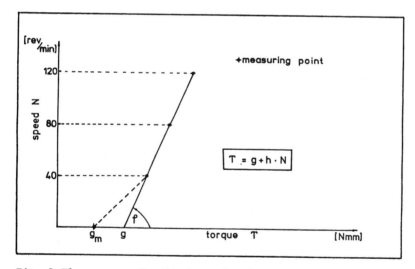

Fig. 3 Flow curve of a Bingham plastic.

The residual 'g_m' is smaller than 'g' which is, according to the Bingham model, an approximation for the yield value. The same results have been reported by Ukraincik (1980) for concrete tested in a shear box. Values for 'g' and 'h' have been calculated from the mean torques at speeds of 40 and 120 rev/min according to the equations (2) and (3).

$$g = T(N=40) - 0.5 * [T(N=120) - T(N=40)] \qquad (2)$$

$$h = [T(N=120) - T(N=40)] / 80 \qquad (3)$$

Banfill found it possible to calibrate the ViscoCorder with precisely defined fluids so that the constants of proportionality relating 'g' and 'h' to yield value and plastic viscosity could be calculated. He determined a mean effective shear rate of $10.2\ N^{1.22}$ (N is the rate of rotation of the sample cylinder (rev/s)). The yield value is given by 7907*'g' (Nmm) and the plastic viscosity by 775*'h' (Nmms). (Banfill: private communication).

3 Results

Fig. 4 shows consistency curves for different types of sand and cement for mixtures of the same composition. Their w/c- and filling-ratios were 0.50 and 0.70, respectively. The value 'g' increases with the fineness of the cement and sand. It is apparent that the cements used in these tests - as opposed to the sands - don't influence 'h' significantly.

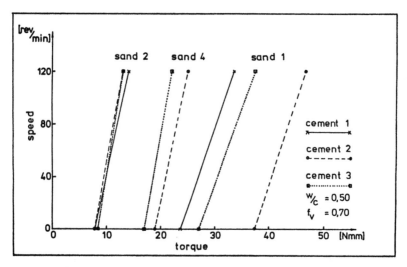

Fig. 4 The effect of different types of sand on the flow curve.

Sand no. 2 (Viscosand) has a considerably smaller proportion of fines below 0,125 mm than sand no. 1. Thus the influence of the different types of cement is smaller because of the reduced interaction between the cement and the sand particles.

The w/c-ratio and filling ratio (f_v) strongly influence the flow curve. Higher filling ratios cause 'g' and 'h' to increase. The sensitivity of a mortar with regard to changes in 'g' and 'h' for different w/c-ratios depends strongly on the actual filling-ratio. A higher w/c-ratio leads to a reduction in 'g' and 'h'.

If the w/c-ratio is larger than 0,65 and the filling ratio is smaller than 0.65 then 'g' becomes so small that it leaves the sensitivity range of the ViscoCorder.

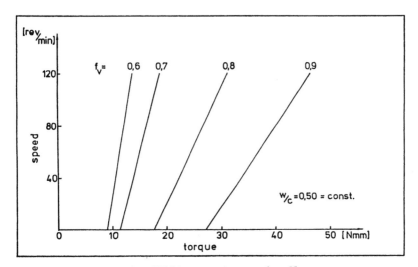

Fig. 5 Effect of the filling ratio on the flow curve.

Figs 9 and 10 show the relation between 'g' determined for mortar-mixtures and the slump and the flow value of corresponding concrete mixtures (cf. 2.5). Further tests are required to find the best composition of the mortar for each particular concrete.

It is apparent in the figures that there is a significant relation between 'g', measured with mortar, and the flow diameter and slump of comparable concretes. A relation between 'h' and these concrete parameters has not yet been found in the present investigations. This may be due to the strong effect of the consistency tests used on the yield value, which masks the effect of the plastic viscosity.

The results show effects comparable to those Tattersall (1987) found in fresh concrete using his Two-Point Workability Apparatus.

Table 3

Influence	Effect on	
	'g'	'h'
Increase of w/c ratio	decrease	decrease
Increase of filling ratio	increase	increase
Increase of fines in the sand	increase	increase or no change
Plasticiser	decrease	no change
Air entraining agent	no change	decrease

233

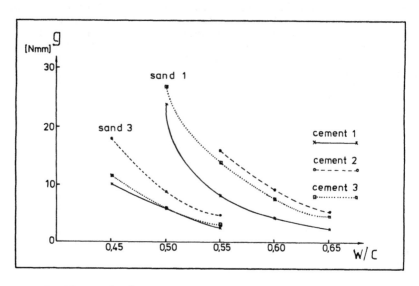

Fig. 6 Effect of w/c-ratio on 'g'.

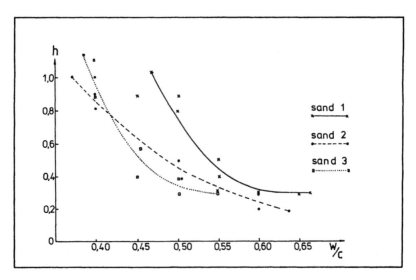

Fig. 7 Effect of w/c-ratio on 'h'.

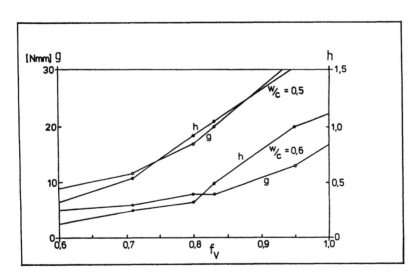

Fig. 8 Effect of filling ratio on 'g' and 'h'.

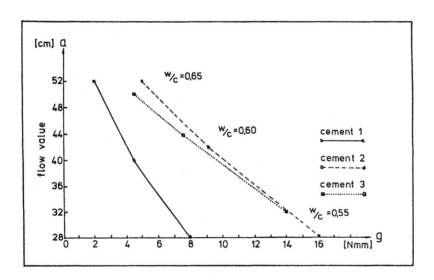

Fig. 9 Effect of 'g' on flow diameter.

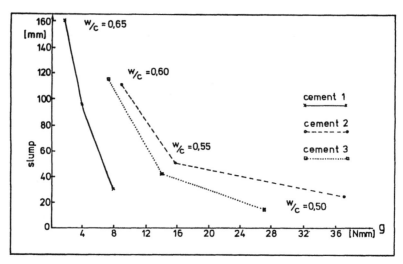

Fig. 10 Effect of 'g' on slump.

4 Conclusions

The two-point test with the ViscoCorder is an appropriate method to
study the influence of mortar components and their composition on the
consistency of mortars.

In preliminary measurements a strong correlation has been found
between 'g', determined for mortar, and the flow diameter and slump of
comparable concretes. However, this result alone is insufficient
since, as is well known, concretes which show the same slump or flow
value can have a very different workability. The essential difficulty
is that the consistency tests in common use give an inadequate description of the actual workability of fresh concrete, whereas other
methods are too costly and/or too difficult to be in widespread application. Furthermore, the effect of the plastic viscosity has not yet
been considered in enough detail. More investigations are necessary to
investigate the correlation between the results obtained with the
ViscoCorder for mortar samples and the consistency of a corresponding
fresh concrete. A quantitative comparison as well as a qualitative description is desirable. The effect of aggregate particles larger than
1 mm should also be considered. At present the ViscoCorder is an effective and simple means for comparing different admixtures, fly
ashes, cement/sand compositions etc. with regard to their qualitative
effect on the workability of fresh concrete. It allows also a qualitative estimate of the stiffening properties of cement in combination
with different sands, fly-ashes and admixtures.

The ViscoCorder allows the determination of the most favourable
combination of w/c-ratio and sand/cement ratio for properties specified by the required workability of the concrete.

The ViscoCorder, produced by Brabender[*], or the model "Viscomat" from Teubert[**] which can be linked to a computer, should be an essential apparatus in every well equipped concrete laboratory. A deeper understanding of the consistency of fresh concrete will contribute to an increase in the quality of concrete structures.

References

Banfill, P.F.G. (1987) The use of a coaxial cylinder viscosimeter   to study the rheology of fresh mortar; **Mitteilungen aus dem Institut für Baustoffkunde und Materialprüfung der Universität Hannover**, Heft 55.

Tattersall, G.H. (1987) Workability measurement and its application to the control of concrete production; **Mitteilungen aus dem Institut für Baustoffkunde und Materialprüfung der Universität Hannover**, Heft 55.

Tattersall, G.H. (1983) Der Zweipunktversuch zur Messung der Verarbeitbarkeit, **Betonwerk+Fertigteiltechnik**, Heft 12, 789ff.

Teubert, J. (1981) Die Messung der Konsistenz von Betonmörteln...; **Betonwerk+Fertigteiltechnik**, Heft 4, S.217-222.

Ukraincik, V. (1980) Study on fresh concrete flow curves; **Cement and Concrete Research**, Vol. 10. pp. 203-212.

Wolter, G. (1985) Messung der relativen Viskosität von Zementmörteln; **Betonwerk+Fertigteiltechnik**, Heft 12, 816-824.

---

[*] Fa. Brabender OHG, Kulturstraße 51-55, D-4100 Duisburg;
[**] Fa. Schleibinger Geräte, Edeltrautstraße 51, D-8000 München 82;

PART SIX

CONCRETES

25 INFLUENCE OF THREE-PHASE STRUCTURE ON THE YIELD STRESS OF FRESH CONCRETE

J. SZWABOWSKI
Silesian Technical University, Gliwice, Poland

Abstract
Yield stress measurements were conducted on a variety of concrete mixes to evaluate the influence of the fresh concrete structure on yield stress and possibility of yield stress prediction when composition is known. The measurements were carried out using rotational viscometer with vane type impeller. For uniform description of three-phase structure of fresh concrete three parameters characerising the structure of aggregate, cement paste ond concrete mix composed of them were proposed. The objectives of the study were relationships between yield stress value and proposed parameters of the structure. It was found that in the absence of normal stresses the yield stress of fresh concrete results from capillary cohesion and varies in broad limits depending on the ratio: liquid dispersing phase volume /solid dispersed phase surface area. This relationship has very distinct maximum corresponding to the constant value of the cement paste volume/ aggregate voids volume ratio. A matematical model of this relationship is given.
Keywords: Fresh Concrete, Rheology, Yield Stress, Structure Parameters, Measurements.

1 Introduction

Rheological behaviour of fresh concrete with regard for three-phase, thick-dispersed structure is complicated. In practice the behaviour of fresh concrete can be represented by the Bingham model. The behaviour is non linear because of abrupt change when yield stress is exceeded. It results from the rheological analysis of mixing, transport, placing and compacting processes that the shear stress/yield stress ratio, determining the possibility of flow, is the first condition of workability. For that reason the study of relation between state of structure and yield stress of fresh concrete is necessery for workability control. Attempts to determine this relation were undertaken by Hobbs (1976) and Bleshchik (1977). Theoretical relations obtained by these authors are dubious both because of the way of analytical description of the state of mix structure and weak agreement with the results of measurements. The aim of this research was to determine in experimental way the relation between yield stress of fresh concrete and its structure, with introduction of variables more suitable for three-phase dispersive structure of the mix.

2 Variables of mix structure

In structural formulation fresh concrete represents three-phase system with high concentration of solids. Identification of the state of mix stucture is based on the formula of composite material composed of the matrix (cement paste) and particles (grains of aggregate). For quantitative description of the structure state this author's concept so: called dispersion factor, presented by author (1975), was used. The concept is based on the statement that from the physical point of view both concrete mix and cement paste and aggregate are dispersed systems. From physical and chemical analysis of such systems follows that the area of interface and filling of system by the continuos disperging phase are of fundamental importance for their properties. In turn, mechanical properties of discussed system depend mainly on links between dispersed particles. For all types of links:contact, capillary and molecular, their strength depends mainly on distances between system particles. For that reason the disperion factor was formulated as the continuous dispersing phase volume V_c /dispersed phase surface S_d ratio

$$D = \frac{V_c}{S_d} \tag{1}$$

Graphical interpretation of the dispersion factor is shown in Fig.1. It should be noted that the factor represents mean thickness of dispersing phase covering the surface of dispersed phase particles or in other words a half of mean

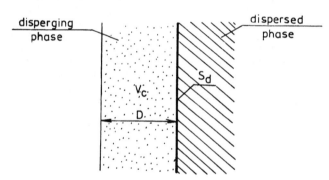

Fig.1. Graphical interpretation of dispersion factor D.

distance between these particles in the system. For above mentioned respects the dispersion factor can be treated as the physical parameter which describes the state of structure of the aggregate, paste and composed concrete mix in a uniform way. For these systems the factor takes on the following form:
 - dispersion factor of aggregate

$$D_a = \frac{V_{vsa}}{S_{sa}} \ (\mu m) \tag{2}$$

where V_{vsa} is specific volume of voids in aggregate and S_{sa} is specific surface of aggregate.

- dispersion factor of cement paste

$$D_p = \frac{V_w}{C \cdot S_{sc}} \; (\mu m) \qquad (3)$$

where V_w is volume of water, C is weight of cement and S_{sc} is specific surface of cement. For given type of cement, when S_{sc} is constant the dispersion factor of the cement paste can be presented in simpler, generally used in concrete technology form

$$D_p = \frac{W}{C} \qquad (4)$$

that, although dimensionless has the same physical meaning.
- dispersion factor of concrete mix

$$D_m = \frac{V_p}{A \, S_{sa}} \; (\mu m) \qquad (5)$$

where V_p is absolute volume of paste and A is weight of aggregate in concrete mix. The presented factors D_a, D_p, D_m, composed set of parameters which author's conviction describes fully the state of fresh concrete structure. D_p is characteristics of the matrix structure, D_a is characteristics of dispersed particles, D_m is characteristics of the composite material. It may be noted that the D_m/D_a ratio is factor of filling of uncompacted aggregate with cement paste.

3 Experimental work

First tests on the relation between the mix structure described as above and its yield stress for Aglite aggregate mixes were desribed by author (1975). The obtained relations presenting the effect of the magnitude of each factor on changes of yield stress are shown in Fig. 2. Limited range of tests made it possible to determine qualitative relations only. The main series of tests were carried out for normal mixes, using sand and coarse aggregate graded with accordance to grading curves presented in Fig. 3. The fine-coarse aggregate ratio by weight was varied for obtaining possibly wide range of the dispersion factor of aggregate. Type one portland cement was used with mean content of $C_3S = 53.2\%$ and Blaine specific surface 3050 cm^2/kg. Because strength and technological requirements determine the value of the W/C ratio. the relations between yield stress τ_o of the mix and the factors D_a, D_m were tested. The tests were divided in two stages. During first stage the changes of τ_o in bifactor space of D_a, D_m were tested. The purpose of the second stage was to determine with greater accuracy relation between τ_o and D_a because in earlier tests the dominant influence of the D_a was found. The tests were carried out for selected crosssection of the factor space determined by $D_m = 60$ μm at 30 levels of D_a factor. In both stages constant magnitude of D_p responding to the 0.5 W/C ratio was used. Measurements of yield stress of the tested mixes were carried out with use of the rotational rheometer developed by this author, described earlier (in 1975). Vane type impeller was used throughout the tests.

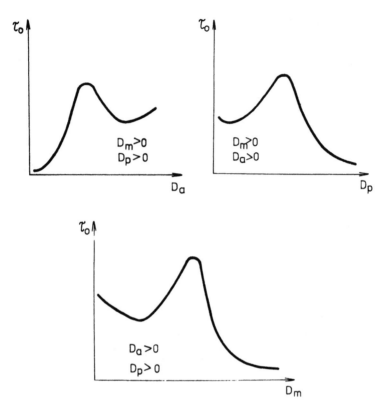

Fig.2. General relationships between yield stress τ_0 of Aglite aggregate mix and parameters of its stucture D_a, D_p, D_m after the author (1975).

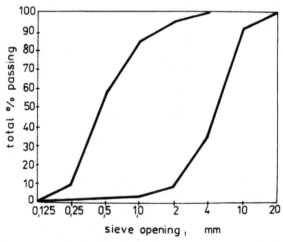

Fig.3. Grading curves of sand and coarse aggregate used in the tests.

Yield stress was measured by the determination of such value of shear stresses in the mix at which the continuous rotational movement of the impeller sunk in the mix begin. The torque acting on the impeller is related with the yield stress by formula

$$\tau_0 = c + \sigma \, \mathrm{tg}\varphi = \frac{T}{2\pi r^2 h} \qquad (6)$$

where: r - radius of impeller, h - height of the impeller working part, T - torque, c - cohesion, σ - normal stress, φ - angle of internal friction. In each point of measurements, in accordance to the tests schedule, three replications were conducted for obtaining τ_0. The mixes were prepared in the pan mixer of batch capacity 150 dm^3.

4 Results and discussion

Obtained results of the relation between τ_0 and D_a and D_m are presented in Fig. 4 and Fig. 5. The shape of curves indicate on very strong relationship between yield stress τ_0 and value of the aggregate dispersion factor D_a and mix dispersion factor D_m. The distinct feature of obtained curves is that their shapes are very similar. The shape of curves $\tau_0(D_m)$ is the mirror reflection of curves $\tau_0(\Delta_a)$ with respect τ_0 the to axis. All curves except $\tau_0(D_a)$ for $D_m = 80$ and 90 μm have distinct zone of maximum values. From discussion of the relation $\tau_0(D_a)$ results

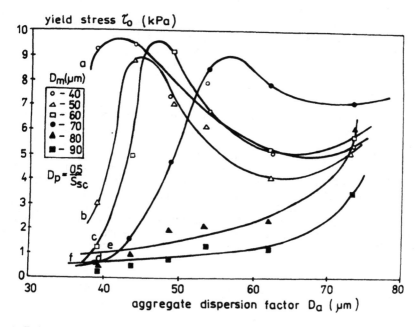

Fig.4. Relationship between concrete mix yield stress τ_0 and aggregate dispersion factor D_a

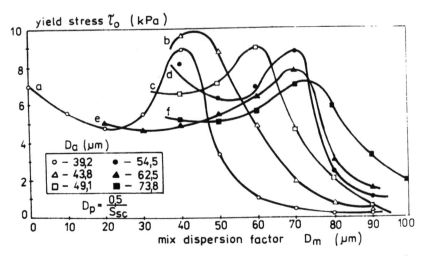

Fig.5. Relationship between concrete mix yield stress τ_o and mix dispersion factor D_m

that the value of D_a for the maximum value of τ_o is rising along with the rise of D_m. In similar way for relation $\tau_o(D_m)$ the value of D_m is rising along with the increase of D_a. Moreover displacement of the maximum is accompanied by displacement of the whole curve. It explains why maxima on curves τ_o (D_a) for D_m = 80 and 90 μm weren't recorded. These curves are only a part of the whole relation $\tau_o(D_a)$ because the zone of maximum due to the mentioned above displace-moment caused by the increase of D_m was outside the tested area. Comparing relations $\tau_o(D_a)$ and $\tau_o(D_m)$ one can note that the increase in aggregate dispersion factor D_a, at constant value of the mix dispersion factor D_m, gives the same effect as decrease of D_m at constant value at D_a. It proves that there exists interrelated influence of D_a and D_m on changes of τ_o. Moreover , it is characteristic that zones of τ_o maximum on particular curves occur at the value of the D_m/D_a ratio equal from 1.0 to 1.2. Presented results of tests on the relation $\tau_o(D_a,D_m)$ indicate that the yield stress of fresh concrete depends mainly on the value of the D_m/D_a ratio . It may be explained by taking into account capillary phenomena. The effect of this phenomena on the value of τ_o may by presented as on Fig. 6. The mix yield stress is the total results of capillary cohesion and internal friction. Because this friction is a monotonic function of aggregate grading and volume concentration, as it was proved by Bombled (1973), extremal course of changes of τ_o can be explained only by changes of capillary cohesion. For $D_m = 0$ (no paste) τ_o corresponds yield stress of the aggregate τ_{oa} and for $D_m \Rightarrow \infty$ τ_o is approachig the paste yield stress τ_{op}. Regression and variance analysis of the data obtained from tests of relation $\tau_o(D_a,D_m)$ made it possible to formulate the mathematical model of this relationship in form

$$\tau_o = \tau_{op} + \tau_{oa} \; \exp\left[a\left(\frac{D_m}{D_a}\right)^2\left(\frac{D_m}{D_a}-b\right)\left(c-\frac{D_m}{D_a}\right)\right] \qquad (7)$$

where a, b, c - constants, depending on the mix type and value of D_m factor. In second stage the influence of the aggregate structure on the mix yield stress τ_o was tested in much wider zone of D_a variability. Obtained results are

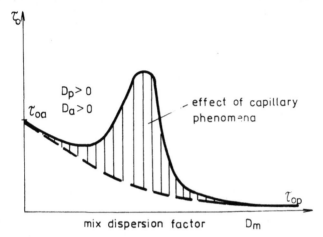

Fig.6. Effect of capillary phenomena on the magnitude of concrete mix yield stress τ_o.

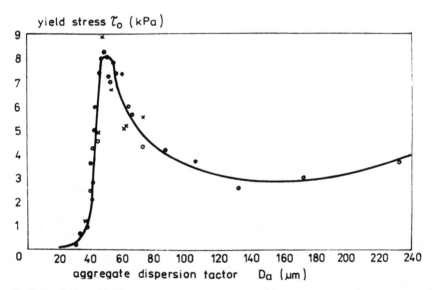

Fig.7. Full relationship between concrete mix yield stress τ_o and aggregate dispersion factor D_a.

presented in Fig. 7. The relation τ_o (D_a) is here very clearly shown by numerous data points. Starting from smallest values, the increase of D_a factor in narrow limit

30 - 50 μm results in very intensive increase of the yield stress, from minimum to maximum value, developed by rapid increase of capillary cohesion. Maximum τ_o takes place at $D_a = 50$ μm which corresponds the 1.2 D_m/D_a ratio. Further increase of D_a causes drop of τ_o at first quick then slower and slower. As the result of regression and variance analysis of the measurement data the mathematical model of the relation $\tau_o(D_a)$, at $D_m = 60$ μm and D_p corresponding to the 0.5 W/C ratio, was obtained in form

$$\tau_o = \tau_{op} + \tau_{oa} \quad \exp\left[\frac{A}{D_a^2}\left(\frac{1}{D_a} - B\right)\left(C - \frac{1}{D_a}\right)\right] \qquad (8)$$

where A, B, C - constants, depending on the mix type and value of D_p and D_m factors.

5 Conclusions

The tests of the influence of three-phase fresh concrete structure on its yield stress made it possible to draw following conlusions: The yield stress of the fresh concrete at its constant temperature depends on the state of the mix structure and its value varies in wide limits. At constant value of the paste dispersion factor D_p the mix yield stress τ_o depends on the value of the aggregate dispersion factor D_a and mix - D_m and more exactly, on the D_m/D_a ratio. Character of the relation $\tau_o(D_a)$ and $\tau_o(D_m)$ and τ_o (D_m/D_a) is extremal one, with very distinct zone of maximum values corresponding to the D_m/D_a ratio within limit from 1.0 to 1.2. Character of relation $\tau_o(D_a)$ and $\tau_o(D_m)$ for coarse aggregate mixes is in full aggrement with the character of the same relation for Aglite aggregate mixes what shows on the possibility to generalize these relations for other types of mixes. The character of the influence of the dispersion factors on the yield stress is in agreement with phenomenological effect of the capillary phenomena in the mix, and in the case of no existing normal stresses the yield stress of fresh concrete depends only on the capillary cohesion.

6 References

Szwabowski, J. (1975) On rheology of concrete mix. part I: On rheological proper ties of concrete mix. **Archiv. Civ.Eng.** vol. 21, 4, (in Polish) pp. 635-648.
Hobbs, D.W. (1976) Influence of Aggregate Volume Concentration upon the Workability of Concrete and Some Predictions from the Viscosity - Elasticity Analogy. **Mag. Concr. Res.** vol. 28, No 97, pp. 191-212.
Bleshchick, N.P. (1977) Structural - mechanical properties and rheology of fresh concrete and pressvacuumed concrete. **Science and Technology.** Minsk (in Russian).
Szwabowski, J. (1975) On rheology of concrete mix, part II: Influence of concrete mix structure on its yield stress. **Archiv. Civ. Eng.** vol 21, 4, (in Polish) pp. 649-665.
Bambled, J.P. (1973) Rheologie des mortiers et des beton frais, in Fresh Concrete, **RILEM Sem . Proc.** vol 2, University of Leeds, pp. 3.1.1 - 3.1.45.

26 MECHANICAL PROPERTIES OF FRESH CONCRETE BEFORE SETTING

L. OULDHAMMOU, P.N. OKOH and Ph. BAUDEAU
Institut Universitaire de Technologie, Department of Civil
Engineering, Strasbourg, France

Abstract
The mechanical property of fresh concrete can be assimila-
ted, under certain conditions, to that of a granular material. The hypothesis advanced here consists in saying that
the forces, subjected to a certain volume of concrete within
a building work, can be simulated and compared to those subjected to a test tube composed of the same material, or of
an equivalent material, in an enclosure of a triaxial cell.
    The evolution of the mechanical behavior of fresh concre-
te was observed in two classical ways; that of a lateral,
constant stress (C.D.) and that of a constant volume (C.U.),
for values of confined stress ranging from 12 to 400 KPa.
For each kind of experimentation, the deviation stress q is
applied with times varying from 0 to 240 minutes, after the
consolidation phase under σ_3 .
    By analogy to certain soils, the fresh concrete is defi-
ned in terms of a characteristic state, which separates two
kinds of rheological behavior : contracting in a subcharacteristic domain, expanding in an overcharacteristic domain.
    The results obtained enabled to highlight the following
points :
- The Luong characteristic frictional angle is constant and
  time independent.
- The Young modulus or initial modulus increases linearly
  with respect to time.
- Cohesion is very small compared to obtained maximum
  stresses.
- Initial Poisson coefficient increases with time.
- The contractancy amplitude decreases with time.
- The pseudo Poisson coefficient of dilatancy decreases
  slightly with time.
Two-path load utilized showed that the drained and
undrained characteristics of the material are dependent; a
result which is already known in soil mechanics and which
applies quite well, in this study, to fresh concrete.
Keywords: Fresh Concrete, Triaxial Test, Micro-Concrete,
Granular Medium, Drained Test, Undrained Test, Deviatoric
Stress, Volumetric Strain, Pore Pressure.

1 Introduction

Fresh concrete is a material that possesses very complex rheological characteristics. It is a heterogeneous material in its composition-sand, gravel, cement. If a certain number of theoretical and experimental studies, Legrand (1971), Bombled (1973), Baudeau (1983),L'Hermite (1948), Barrioulet, Legrand (1978), have been carried out so far, none of these studies can forecast the mecanical behavior of fresh concrete under diverse loadings. Based on a precedent study, Baudeau (1983), in which fresh concrete was considered as an essentially granular medium and for which a certain number of parameters defining its mechanical behavior were highlighted, this article will consider the influence of time factor on the evolution of these parameters as well as the rheological behavior of fresh concrete in triaxial tests.

2 Characteristic parameters of fresh concrete

Previous experimental results, Baudeau (1983) and Darve (1978) have enabled to define , from stress-axial and volumetric strain curves, parameters that can be used to characterize the mechanical behavior of fresh concrete Fig. 1
Where: ①— Internal friction coefficient at plasticity
　　　　landing: ϕ.
　　　②- Cohesion: C.
　　　③-Internal friction characteristic: ϕ_c.
　　　④-Young modulus: Ui.
　　　⑤-Initial Poisson ratio: $\mathcal{V}$.
　　　⑥ - Amplitude of contractancy: $\mathcal{E}_{vc}$.
　　　⑦ -Pseudo Poisson ratio: $\mathcal{V}_m$.

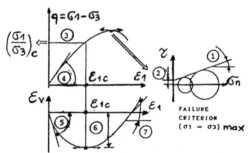

Fig.1. Constitutive characteristics.

3 Experimental procedure and material used

The concrete is in reality a micro concrete composed of four aggreagate classes: 0/0.315; 0.315/0.63; 0.63/1; 1/2. The water-cement ratio w/c is equal to 0,46. The material is slightly vibrated while placing it in triaxial cells of diameter = 70 mm and height = 140 mm. Two-path load tests were used:
- that of constant lateral stress: consolidated, drained

test (C.D.), and
- that of constant volume: consolidated, undrained test
(C.U.). For each type of test, the deviation stress
$q = \sigma_1 - \sigma_3$ is applied in times varying from 0 to 240 minutes,
after having consolidated under σ_3. The applied lateral
stresses are as follows: 12, 50, 90, 130, 200, 250 and 400 KPa.
These pressures were applied at a constant rate during the
period of testing.

4 Experimental results of consolidated, drained tests (C.D)

The stress-axial and volumetric strain curves enabled to
show 7 characteristic parameters see Fig. 1. These parameters were studied in function of the lapse of experimental
time spent only after the consolidation phase under σ_3.

4.1 Curves of stress versus axial strain $(\sigma - \varepsilon_1)$
Fig. 2 represents the evolution of the deviation stress q
in function of the relative axial strain ε_1, for times
varying from 0 to 240 minutes and for $\sigma_3 = 250$ KPa. Results
show a change in initial slopes representing the Young
modulus, as well as a difference between the peaks of stress
that materialize the start of plastic flow of the material.
Fig. 3 show that the maximum stress/confined stress is an
increasing function of time.

- Internal friction coefficient at plasticity landing: ϕ ①
The failure criterion that was used corresponds to the maximum value of the total axial stress σ_1 max and supposes that
fresh concrete responds to the Coulomb law. By observing
Fig. 4, one can see that $tg\phi$ increases slightly with time.
the reduction of the water content during testing loads to
lessening the lubricant effect of the cement paste with
respect to the aggregates. More over, the setting phenomenon
of the binding material begins to become important. The same
remarks have been made by Barrioulet and Legrand (1977),
(1978) in the study of fresh concrete flow in a
maniabilimeter. It is necessary to precise here that the
linear form of the Coulomb law is only but an approximation
in the approach to the mechanical behavior of the material.

- Cohesion: C ②
The Mohr circle envelope obtained during triaxial tests is a
straight line that enables to definie the cohesion of a
material. The results of Fig. 5 show that cohesion increases
in function of time; however, it remains negligible before
the maximum values of the total stresses. The Mohr-Coulomb
and Rowe theories (1962) and the studies realized by
Alexandridis (1979) and Alexandridis, Gardner (1981) on the
shearing of fresh concrete, were used to confirm our results.

- Internal friction characteristic: ϕ_c ③
The internal friction characteristic given by:

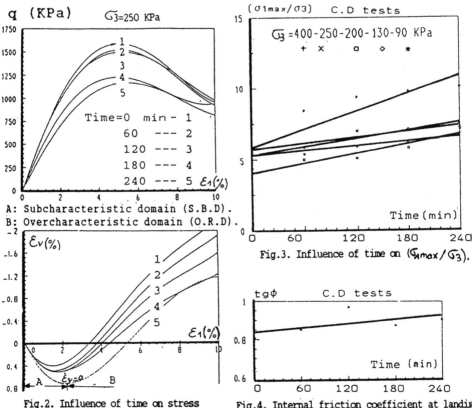

A: Subcharacteristic domain (S.B.D).
B: Overcharacteristic domain (O.R.D).

Fig.3. Influence of time on (σ_{1max}/σ_3).

Fig.2. Influence of time on stress
deviation and volumetric strain.

Fig.4. Internal friction coefficient at landing
of plasticity in function of time.

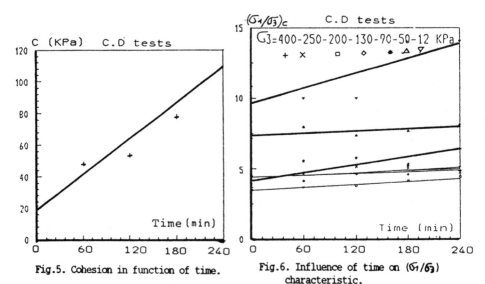

Fig.5. Cohesion in function of time.

Fig.6. Influence of time on (σ_1/σ_3)
characteristic.

$$(\sigma_1/\sigma_3) = tg^2(\frac{\pi}{4} + \frac{\phi c}{2})$$ (1)

was called " aggregate friction " by Kirkpatrick (1961).
This relationship characterizes the capacity of entanglement
of fresh concrete. After analyzing the experimental results
of Fig. 6, it seems that the internal friction characteristic
is constant for $\sigma_3 \geqslant 130$ KPa and variable for $\sigma_3 \leqslant 90$ KPa,
but increases with time. this is quite normal, since the
fresh concrete begins to harder with time.

Remarks:
- The way of determining the point where the volume deformation is equal to zero, according to the curve $\Delta V/V_0$
(ΔV: volume variation of the sample), is not precise since V_0
the initial volume, is obtained from the geometrical
dimensions of the sample which is considered in theory as a
cylinder. In reality, it is different. This reality can
influence the exact position of $\mathcal{E}v = 0$ see fig. 7. That is
why we use, in our calculations, $\Delta V = 0$.
- For what concerns the determination of the critical state
defined by Habib and Luong (1978), it seems that it is
preferable to take into consideration the variation rate of
$\dot{P}=0$ applied axially on the test-sample instead of the
variation rate of $\dot{q} = 0$, since here also there is the
influence of the measure precision of the test-sample's
section to consider, see Fig. 8.

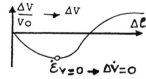
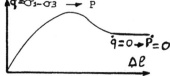

Fig.7. Volume variation versus   Fig.8. Load variation versus
    absolute axial strain.   absolute axial strain.

- No matter the time and σ_3, except for $t \geqslant 180$ min and
$\sigma_3 = 12$ KPa, fresh concrete is defined by a characteristic
state which differentiates two types of rheological behavior:
  a) Contractancy of the subcharacteristic domain,
  b) Dilatancy of the overcharacteristic domain.
- The rheological existence of the stress peak is actually
being put to question. Dresher and Vardoulakis (1982) studied
soil mechanics tests in view of the theory of bifurcation.
Desrues (1984) and Duthilleul (1983), by the stereophoto-
grametry method, brought to evidence the localization of
deformations in biaxial tests. These measures indicate, for
this type of tests,that the localization appears very early,
even before the peak. from the whole of our results, we
noticed that the failure appears before the peak and then
propagates to yield a complet mechanism of shear bands and
blocks. This failure control was experimented on a completely
dilating material with an imposed surface of failure. By
the help of volume variation measures in the cell and the
pore volume, we were able to localize the failure before the

peak. This last important point will be treated in detail in a next publication.

- Young modulus: Ui ④
The Young modulus is the limit of $(d\sigma/d\varepsilon_1)$, when $d\varepsilon_1 \to 0$. According to Fig. 9, Ui increases linearly in function of time. The difference between the various straight lines is due to the influence of σ_3 on their points of departure.

4.2 Curves of volumetric strains versus axial strains: $\varepsilon_v - \varepsilon_1$
An analysis of Fig. 10 shows that:
 a) at a constant ε_1, the volumetric strain is a decreasing function of time.
 b) ε_v is an increasing function of q, until the point where the volume variation rate is equal to zero: $\dot{\varepsilon}_v = 0$.
 c) the two curves that are parallel to ε_v and which go by the point A of $\dot{\varepsilon}_v = 0$ of the curve t = 0 min and the point B of the curve t = 240 min, form up the variation of the point ε_v in function of ε_v. The same figure gives information on the maximum value of the volumetric strain and the maximum stress at the plasticity landing.

- Initial Poisson ratio: ν_i ⑤
The elastic lateral strain is characterized by the initial slope of the curve ε_v in function of ε_1. This slope is function of the initial Poisson ratio of the material. According to experimental results on Fig. 11, the influence of σ_3 on ν_i is important with respect to time. This same kind of phenomenor can be seen in the case of sands under high values of σ_3, according to the experiments of Habib and Luong (1978). A study was realized by Baudeau (1985), which shows quite well the influence of σ_3 on the Poisson ratio in the case of fresh concrete (see Fig. 12).

- Amplitude of contractancy: ε_{vc} ⑥
The compacity of fresh concrete increases with time, because the volume of expelled water is quite important. This results to a diminution of the contractancy, see Fig. 13 and an increase of the resistance to shear. At a slight value of confinement stress $\sigma_3 = 12$ KPa and for a time $t \geqslant 180$ min, the loading leads only to a volume expansion until a plastic flow of the concrete, while for $\sigma_3 \geqslant 12$ KPa, the loading leads to a volume contractancy followed by a continuous expansion until the flow of the material. Fresh concrete is considered here as a granular material having a tendency of presenting more of a contractancy behavior than that of dilatancy, so long as the confinement stress σ_3 continues to increase. This was confirmed by the experimentations of Luong (1978) which show that for sands, the dilatancy disappears for high values of σ_3.

- Pseudo Poisson ratio: ν_m ⑦
ν_m is characterized by the slope of volume variation curve,

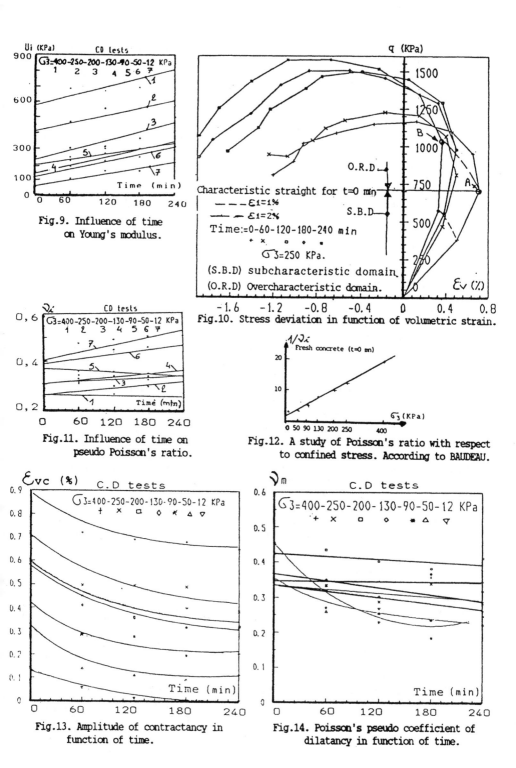

Fig.9. Influence of time on Young's modulus.

Characteristic straight for t=0 min
−− − −ει=1%
−·− − ει=2%
Time:=0-60-120-180-240 min
+ × □ ◆ ●
σ3=250 KPa.
(S.B.D) subcharacteristic domain.
(O.R.D) Overcharacteristic domain.

Fig.10. Stress deviation in function of volumetric strain.

Fig.11. Influence of time on pseudo Poisson's ratio.

Fig.12. A study of Poisson's ratio with respect to confined stress. According to BAUDEAU.

Fig.13. Amplitude of contractancy in function of time.

Fig.14. Poisson's pseudo coefficient of dilatancy in function of time.

255

in the overcharacteristic domain. Fig. 14 shows a slight
influence of time on ϑ_m, for $\sigma_3 \gg 200$ KPa. The slight decline
of ϑ_m in function of time is due to the fact that the slopes
of $\mathcal{E}_v$ increases with time, since ϑ_m is calculated from:

$$1 - 2.\vartheta_m = \mathcal{E}_v / \mathcal{E}_1 \qquad\qquad (2)$$

Also, it is known that fresh concrete has a more tendency of
dilating than contracting with age, according to Fig. 2. The
contractancy increases, the more ϑ_m approaches 0.5; then its
slope becomes equal to zero. There are two possibilities
where there can be no dilatancy:
   a) the case of a loose material;
   b) the case of a very high confinement stress σ_3.

5 Experimental results of consolidated undrained tests (C.U)

The experimental results of the characteristics parameters
of both C.U and C.D tests are not independent.

- Pore pressure versus axial strain curves: (U-$\mathcal{E}_1$)
The tendency to contraction or expansion of fresh concrete
will be compensated by an increase or a diminution of the
pore pressure. In the case where its rate of variation is
equal to zero, the value of its axial strain corresponds to
that of $\mathcal{E}_v = 0$. The appearance of the maximum on the evolution of the pore pressure corresponds to the ratio of the
principal effective stresses that define the internal
friction characteristic. Fig.15 indicates the influence of
time on the maximum pore pressures. The pore pressures
decrease as from t = 0 min until t = 60 min, then increase
linearly until t = 240 min. This is due to the fact that
there subsists a little amount of water in the concrete;
however, pore pressure increases as σ_3 increases.
When Fig. 16 was analyzed, it was seen that:
   a) at a constant time, U increases in function of $(\sigma_1 - \sigma_3)$,
then decreases as from the point where the variation rate of
the pore pressure becomes null $\dot{U} = 0$. This point separates

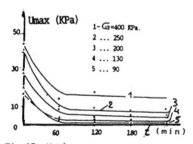

Fig.15. Maximum pore pressure versus time.

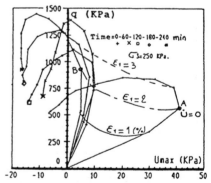

Fig.16. Stress deviation in function
of pore pressure.

the two subcharacteristic domains (S.B.D) from the
entanglement process and the overcharacteristic domain
(O.R.D) from the disentanglement process of grains;
  b) at a constant pore pressure, q=f(t) decreases with
respect to $\mathcal{E}_1$(as from $\mathcal{E}_1$= 3 %); this helps to obtain
information on the failure;
  c) the two lines that are parallel to U and which go
through the point A, from $\mathring{U}$ = 0 at t = 0 min, and the
point B, at t = 240 min, represent the variation of the
characteristic domain in function of time.

6 Conclusion

The results derived from the various tests which enabled to
determine the characteristic parameters of fresh concrete
subjected to biaxial compression forces in drained and
undrained conditions, with confinement stresses ranging from
12 to 400 KPa and times after consolidation ranging from 0 to
240 minutes, permit to formulate the following conclusions:
. 1) Fresh concrete is defined by a characteristic state of
$\mathcal{E}v$= 0 or $\mathring{U}$ = 0, which separates two kinds of rheological
behavior: contractancy in the subcharacteristic domain
(S.B.D) and dilatancy in the overcharacteristic
domain (O.R.D).
  2) Both σ_1-σ_3 and the Young modulus increase linearly in
function of time.
  3) The internal friction coefficient at plasticity landing
Φ and cohesion increase linearly with time. Cohesion remains
negligible before qmax
  4) The internal friction characteristic is independent of
time, for σ_3 ⩾ 130 KPa, but variable for σ_3 < 130 KPa.
  5) The initial Poisson ratio only increases
slightly in function of time.
  6) For little values of σ_3
and times ⩾ 180 min, the
contractancy amplitude gives
way to dilatancy. Based on
the studies of Luong (1978)
on granular materials like
sand, it can be said that
time and σ_3 tip up certain
characteristic parameters
of fresh concrete from a
minimum to a maximum

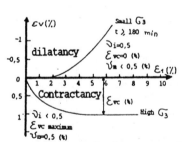

Consequently, this leads
to a change in rheological
behavior, Fig. 17.

Fig.17. Influence of confined stress on rheological
behaviour for fresh concrete.

  From our experimental results, time and σ_3 can be divided
into two groups:
  a) σ_3 ⩾ 90 KPa and t ⩾ 60 min, during which the characteristic parameters appear to be constant;
  b) σ_3 < 90 KPa and t < 60 min: sensitive parameters.
  The mechanical properties of fresh concrete are variable

with respect to the age of the material and to the forces
applied on it. An account of these properties, in the
comparative study of a concrete and a micro concrete, will
be the subject of a next publication.

7 References

Legrand, C. (1971) Contribution à l'étude de la rhéologie
du béton frais. **Ph.D. Thesis,** University of Toulouse.
Bombled, J.P. (1973) Rhéologie des mortiers et des bétons
frais, influence du facteur ciment. RILEM, Paris.
Baudeau, Ph. (1983) Contribution à l'étude du comportement
rhéologique du béton frais contenu dans des parois
coffrantes de grande hauteur avant prise du ciment.
Ph.D. Thesis, University of Strasbourg.
L'Hermite, R. (1948) Memoires sur la mecanique physique du
béton. **Annales I.T.B.T.P.**
Barrioulet, M. and Legrand, C. (1978) Étude des frottements
intergranulaires dans le béton frais. Idées nouvelles sur
l'écoulement du béton frais vibré. **Revue des Matériaux et
Constructions,** N° 63, pp. 191-197.
Darve, F. (1978) Une formulation incrémentale des lois
rhéologiques. Application aux sols. **Ph.D. Thesis,** Grenoble.
Barrioulet, M. and Legrand, C. (1977) Influence respectives
de la pâte et des granulats sur l'aptitude à l'écoulement
du béton frais. Rôle joué par les granulats. **Revue des
Matériaux et Constructions,** N° 60.
Rowe, P.W. (1962) The stress-dilatancy relation for static
equilibrium of an assembly of particles in contact.
Proceedings of the Royal Society, series A, Vol. 269.
Alexandridis, A. (1979) Mechanical properties of fresh
concrete. **M.A.Sc. Thesis,** University of Ottawa.
Alexandridis, A. and Gardner, N.J. (1981) Mechanical
behaviour of fresh concrete. **Cement and Concrete
Research,** Vol. 11, pp. 323-329, USA.
Kirkpatrick, W.M. (1961) Discussion on soil properties and
their measurement. **Proc, 5 th International Conference on
Soils Mechanics,** Vol. III, pp. 131-133, Paris.
Habib, P. and Luong, M.P. (1978) Sols pulvérulents sous
chargement cyclique. **Séminaire, Matériaux et Structures
sous Chargement Cyclique,** Ecole Polytechnique, Paris.
Dresher, A. and Vardoulakis, I. (1982) Geometric softening
in triaxial test on granular material. **Revue Française
de Géotechnique,** N° 4, pp. 291-303.
Desrues, J. (1984) La localisation de la déformations dans
les matériaux granulaires. **Ph.D. Thesis,** Grenoble.
Duthilleul, B. (1983) Rupture progressive-simulation physique
et numérique. **Ph.D. Thesis,** University of Grenoble.
Baudeau, Ph. (1985) Rheological study on fresh micro concrete.
Determination of characteristic parameters. **4 th National
Conference with International participation on mechanics
and technology of composite materiels,** Varna, Bulgaria.

7 EXPERIENCES TO MEASURE THE WORKABILITY OF NO-SLUMP CONCRETE

K.J. JUVAS
Partek Corporation, Pargas, Finland

Abstract
This paper deals with a study, the purpose of which was to
find out the factors affecting the workability of no-slump
concrete. A new apparatus called Intensive Compaction
Tester has made tests possible. Even small changes in
workability can be observed by the tester. Investigations
show that the workability is improved by water, super-
lasticizer, silica fume, high-grade sand and round-shaped
gravel. The workability is deteriorated by poor sand,
crushed aggregate, elevated temperature and a long delay
before casting.
Keywords: No-slump concrete, Workability, Testing, Silica
fume, Aggregate, Hollow-core slab, Superplasticizer

1 Introduction

The workability of no-slump concrete has been investigated
very little. Lack of appropriate methods and equipment has
been an obstacle to the research work. Presently almost
all methods for testing workability are designed for more
flowable concretes.
Because of the difficult workability of no-slump concrete, the quality control of the concrete mix is very
important. The smallest changes in the raw material cause
so great changes in the workability that the quality of the
finished products changes. If the no-slump concrete is
properly compacted, the properties of the finished product
are excellent.
An example and perhaps the most common application for
no-slump concrete is prestressed, extruded hollow-core
slabs. The valuable properties of no-slump concrete can
all be utilized in the manufacture of hollow-core slabs:

Good stability before hardening.
High early and grading strength by an appropriate
amount of cement.
The good durability and small deformations of hardened
concrete.

In this study a new test method including an equipment called Intensive Compaction Tester (ICT) has been used for testing no-slump concrete for hollow-core slabs. The tester is designed specifically for testing the properties of no-slump concrete. The tester can be used for monitoring workability and for the manufacture of cylindrical samples to test the compressive strength. The tests have been carried out by Partek Corporation's Concrete laboratory at Pargas, Finland.

2 Test method

The tester used in this study to measure workability has a control unit and a main unit. A balance is also needed. The tester's different parts are shown in Figure 1.

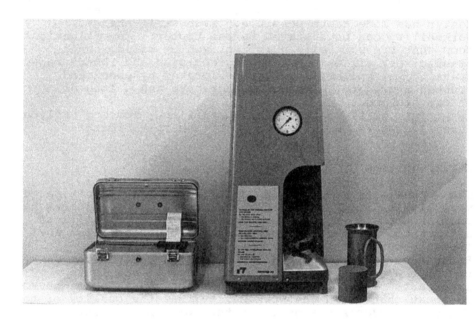

Fig. 1. Intensive Compaction Tester.
Control unit, left, main unit, centre, and the work cylinder and a compacted sample, right.

The different stages of the workability test are shown in Figure 2. The material to be tested is placed in the work cylinder. Its weight is recorded. The normal amount of sample is 900 g. Test parameters are entered into the control unit. The compacting pressure and the number of compaction cycles can vary, normally ranging from 0.2 to 0.4 MPa (2 to 4 bar) and 80 to 160 cycles, respectively.

The material sample is compacted using the intensive compaction method according to the instructions preprogrammed into the control unit. A mass sample is pressed between top and bottom plates in the work cylinder. The top and bottom plates remain parallel, but the angle between the plates and the work cylinder changes constantly during the circular work motion. The material moves along shear planes as it is worked. The shearing motion together with the axial compressive force allows particles to realign into more favourable positions and air is forced from the sample. The tester continually measures the height of the concrete cylinder. During testing the control unit continues to calculate density on the basis of the height and the known weight of the sample.

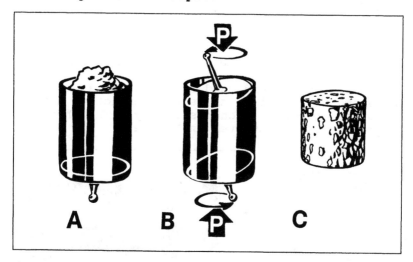

Fig. 2. Different stages of the workability test.

A)   The weighed sample in the work cylinder.
B)   The sample is compacted with the aid of circular motion and compressive force. In the figure the angle between the plates is made too great to illustrate the operating principle of the cylinder.
C)   The sample after it has been removed from the test cylinder.

After completed compaction the control unit prints out a test report. The sample's density is printed out after 10, 20, 40, 80, 160 etc. compaction cycles. Often 80 or 160 cycles are sufficient to draw a curve illustrating the compaction. Figure 3 gives an example of the test report and the compaction curve. The sample is removed from the work cylinder and can be used for further testing, e.g. strength testing.

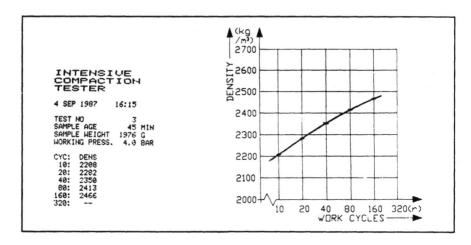

Fig. 3.   Test report and compaction diagram.

This compaction method used in the test has been found to correspond very well to the shear compaction method modern hollow-core slab machines employ. The shear compaction method is illustrated in Figure 4.

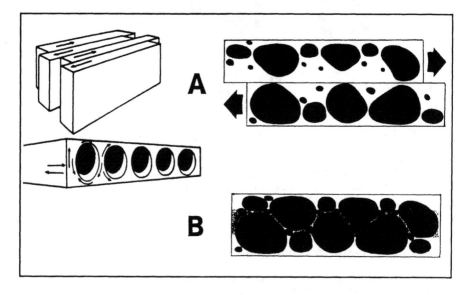

Fig. 4.   Shear compaction method developed by Elematic
          Engineering for hollow-core slabs.
          The aggregates are brought into appropriate
          positions during shear compaction.
          A) Concrete during shear compaction.
          B) Compacted concrete.

3 Tests

3.1 Proportioning

The effect of some of the most common material variables on the workability of no-slump concrete was tested. The variables included water amount, superplasticizer content, silica fume, grade of sand and gravel, the temperature of fresh concrete and the age of concrete during testing. Table 1 shows the proportioning used in the tests.

Table 1.   Tested proportioning.

Con-crete No.	Sand A kg/m³	Sand B kg/m³	Gravel A kg/m³	Gravel B kg/m³	Silica fume kg/m³	Super-plastic. kg/m³	Water /w/c kg/m³
1	1350	-	725	-	-	-	110/0.33
2	"	-	"	-	-	-	105/0.30
3	"	-	"	-	-	-	115/0.35
4	"	-	"	-	-	3.3(1%)	110/0.33
5	"	-	"	-	-	6.6(2%)	"
6	"	-	"	-	26(8%)	-	"
7	"	-	"	-	26(8%)	6.6(2%)	"
8	-	1350	"	-	-	-	"
9	-	1350	"	-	-	3.3(1%)	"
10	-	1350(w)	"	-	-	-	"
11	1350	-	310	415	-	-	"
12	1245	-	-	830	-	-	"
13	830	-	-	1245	-	-	"

Rapid-hardening Finnish portland cement with a specific surface (Blaine) of about 480 m²/kg was used in all concretes. All aggregate types were Finnish granitic rock. Sand A is washed sand normally used in laboratory tests. Its silt content, as measured by the sedimentation method, is not higher than 1%. Sand B is sieved, not crushed aggregate with a silt content of 6%. Concrete No. 10 contains sand B, which has been washed to reduce the silt content to 2.5% (marked 1350,w). Gravel A is washed natural gravel normally used in laboratory tests. Gravel B is crushed, unwashed gravel. The combined aggregate grading is continuous, i.e. efforts have been made to include all grain sizes equally. The silica fume was Norwegian origin. Modified naphthalene was used as superplasticizer.

3.2 Tests performed

Concrete mixes were made according to the above proportionings. Their temperature was +22° ± 2°C after mixing. Immediately after mixing the workability of each concrete mix was tested by the Intensive Compaction Tester. The test was repeated six times. The time required for the workability test was 30 minutes per concrete mix.

Moreover, proportioning No. 1 was used to test +30°C and +60°C warm concretes and +20°C concrete at an initial age of 30 and 60 minutes.

A compressive force of 0.4 MPa (4 bar) and 80 compression cycles were employed in the test. These values have been found to give the same rate of compaction as a hollow-core slab machine.

4 Results

Test results obtained with +20°C concretes are shown in Table 2 as mean values of six tests. The results showed a very small deviation, 7 kg/m³ on an average. The mixes stiffened slightly during the 30 minute test, the average difference between the first and sixth test being 15 kg/m³. The results obtained at an age of 45 and 60 minutes and at higher temperatures are shown in Table 3. These concretes, mainly due to elevated temperature, showed a more significant stiffening during the test. The obtained results are also illustrated in Figure 5 to 10.

Table 2.   Workability test results.

Concrete No.	Density of cylinder sample (kg/m³) Compression cycles		
	20	40	80
1	2232	2321	2395
2	2177	2280	2340
3	2273	2368	2430
4	2247	2342	2435
5	2285	2405	2520
6	2265	2358	2442
7	2305	2415	2535
8	2225	2309	2363
9	2211	2295	2332
10	2230	2315	2383
11	2230	2310	2382
12	2215	2285	2352
13	2190	2253	2305

Table 3.   Workability of concretes of different ages and temperatures, proportioning No. 1

Age at the beginning of test (min)	Density of cylinder sample (kg/m³) Concrete temperature (°C)		
	20	30	50
5	2395	2390	2383
30	2380	2367	2329
60	2367	2332	2260

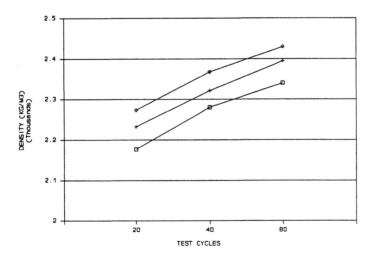

Fig. 5. The effect of water amount on workability.
Legend:
- + - concrete No. 1, 110 l/m^3, w/c = 0.33
- □ - concrete No. 2, 105 l/m^3, w/c = 0.30
- ◇ - concrete No. 3, 115 l/m^3, w/c = 0.35

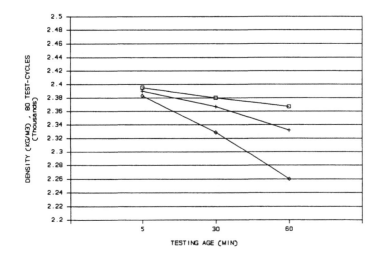

Fig. 6. The effect of concrete temperature and age on workability. Mixture proportioning No. 1.
Legend: concrete temperature
- □ - +20°C,
- + - +30°C,
- ◇ - +50°C.

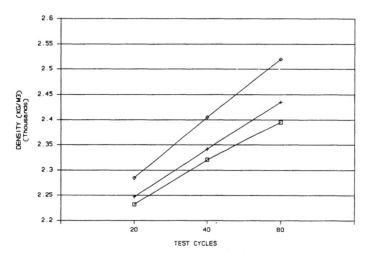

Fig. 7. The effect of superplasticizer on workability
Legend:
- □ - concrete No. 1, 0% of superplasticizer
- + - concrete No. 4, 1% of superplasticizer
- ◊ - concrete No. 5, 2% of superplasticizer

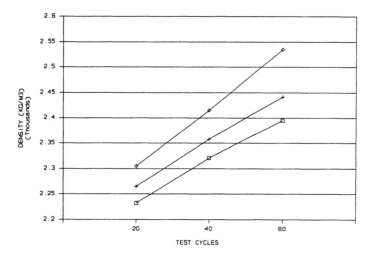

Fig. 8. The effect of silica fume on workability.
Legend:
- □ - concrete No. 1, 0% of silica fume
- + - concrete No. 6, 8% of silica fume
- ◊ - concrete No. 7, 8% of silica fume
       and 2% of superplasticizer.

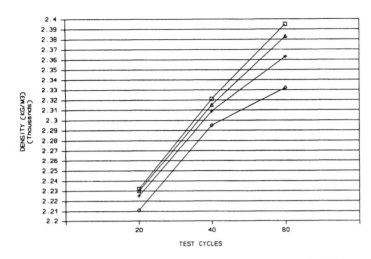

Fig. 9. The effect of aggregate on workability.
Legend:
- □ - concrete No. 1, 1% of silt in aggregate
- ◇ - concrete No. 8, 6% of silt in aggregate
- Δ - concrete No. 9, 6% of silt and 1% of
       superplasticizer,
- + - concrete No. 10, 2.5% of silt (washed
       aggregate).

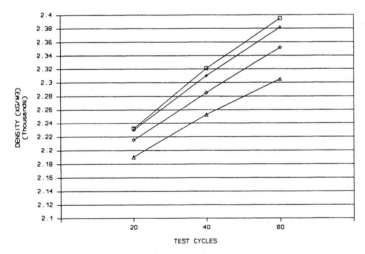

Fig.10. The effect of crushed aggregate on workability
Legend:
- □ - concrete No. 1, 0% of crushed aggregate
- + - concrete No. 11, 20% of "   "
- ◇ - concrete No. 12, 40% of "   "
- Δ - concrete No. 13, 60% of "   "

5 Conclusions

5.1 Test method

The Intensive Compaction Tester (ICT) used for determining workability has proved to be easy to operate, fast and reliable. The sample is compacted automatically. Thus the human influence on the test result is very small. One test takes about 5 minutes for an experienced user. The standard deviation of six successive tests was 7 kg/m³, mainly due to mixture stiffening during testing. Even small changes in the proportioning of no-slump concrete can be observed by the tester, such as a 5 l/m³ change in the water amount and a 2% increase in the silt content of aggregate. The compacted cylinders can be used later to test the compressive strength. Repeated tests made on the same concretes show that the test has a good repeatability.

5.2 Test results

The workability of a concrete with a water-cement ratio of 0.30 can clearly be improved by increasing its water amount by 5 to 10 litres. The improvement is 55 to 90 kg/m³ as measured by the IC Tester. In practice an increase in the water amount is limited by deteriorating concrete stability, which is of crucial importance in the hollow-core slab production.

The workability of no-slump concrete deteriorates as a function of temperature and time. At a temperature of +20°C the deterioration is only 15 kg/m³ after 30 minutes. When the concrete temperature is raised to +50°C, the deterioration is 54 kg/m³ at an age of 30 minutes, and the concrete is not suitable for hollow-core slab manufacture.

A 1 to 2% dosage of superplasticizer has a clearly improving effect on the workability. The improvement is as high as 40 to 125 kg/m³. Practical experience shows that the use of superplasticizer reduces the stability of concrete less than water does.

Addition of silica fume also improves the workability of no-slump concrete. A silica fume content of 26 kg/m³ (8%) improves the workability by 47 kg/m³. When both silica fume and superplasticizer are used, the improvement is 140 kg/m³.

An increasing content of silt in aggregate has a reducing effect on the workability. An increase of silt from 1% to 6% reduces the workability by 63 kg/m³. This can be prevented by washing the aggregate (concrete 10) or by using a superplasticizer (concrete 9).

Crushed aggregate deteriorates the workability of no-slump concrete. The effect is only 13 kg/m³, when the amount of crushed grains is 20% of the total amount of aggregate. When the amount is raised up to 40% and 60%, the effect is clearly higher: 43 kg/m³ (40%) and 90 kg/m³ (60%).

6 References

Juvas, K.J. (1987) The workability of no-slump concrete
   for use in hollow-core slabs. **Nordic Concrete Research**
   No. 6, The Nordic Concrete Federation, Oslo,
   pp. 121-130
Paakkinen, I. (1986) Intensive Compaction Tester. **Nordic
   Concrete Research** No. 5, The Nordic Concrete Federation, Oslo, pp. 109-116.
Schwartz, S. (1984) Practical hollow-core floor slab
   production below 85 dB(A). **Beton+Fertigteil-Technik**
   12/84, pp. 807-813.

28 APPLICATION OF RHEOLOGICAL MEASUREMENTS TO PRACTICAL CONTROL OF CONCRETE

G.H. TATTERSALL
Department of Civil and Structural Engineering, University
of Sheffield, UK

Abstract
The fact that the flow properties of fresh concrete
approximate closely to the Bingham model, over the range
of shear rates important in practice, explains why it is
necessary to consider workability in terms of not fewer
than two constants, and why any single-point test is
inadequate even for pass/fail use.
    In addition to its ability to differentiate correctly
between concretes wrongly adjudged identical by a single-
point test, the two-point test has the important
advantage that its results can be used to identify
factors causing unwanted variability of concrete, by
studying the nature of changes in g (yield value) and h
(plastic viscosity), and the relationships between them.
Practical application of these principles on several
sites is described.
    A method of establishing the best combinations of g
and h for a practical job, in terms of an apparent
viscosity at a postulated effective shear rate, is
outlined and illustrated by a practical application to
the case of production of hydraulically pressed slabs.

Keywords: Bingham model, Quality control, Site
          investigations.

1   Introduction.

The well established fact that, at the low shear rates
important in practice, the flow properties of fresh
concrete approximate closely to the Bingham model, has
important consequences for the control of concrete
production by means of workability measurement. Since
the material is characterised by two constants, the yield
value and the plastic viscosity, and since there is in
general no correlation between the values of the two, it
is clear that any test in which a measurement is made at
only one shear rate, or under one set of shearing
conditions, is incapable of providing sufficient
information. This criticism applies to all the tests
that are incorporated in British and other national
Standards and means that they cannot even operate
satisfactorily to provide pass/fail criteria. Practical
experience on site bears out this statement in that it is

270

ell known that, for example, two concretes of the same lump may behave quite differently on the job.

Measurement of the two Bingham constants permits ready differentiation of concretes that may be wrongly adjudged identical by one of the Standard tests and so provides a facility that is essential for the development of even the most rudimentary control system. In addition, there is a further advantage arising from the fact that the various factors in mix composition that affect workability, affect yield value and plastic viscosity in different ways, so a study of the nature of the changes that occur can provide information to indicate what is the causative factor. The various effects may be summarised, with some simplification for practical purposes in making site decisions, as in Table I.

TABLE I   Factors Affecting Workability

Cause	Effect on	
	Yield Value	Plastic Viscosity
Increase water	Decrease	Decrease
Increase plasticiser	Decrease	None
Increase A.E.A.	None	Decrease
Change Cement	Depends on particular mix	
Change fines	Depends on particular mix	

It will be noticed that the first three factors have effects which, at least in direction, (i.e. increase or decrease) are the same for all mixes, whereas the others have effects whose size <u>and</u> direction depend on the mix under consideration. In other words, for the first three there are no interaction terms but for the rest there are. Thus, increase in water content always decreases both yield value and plastic viscosity whereas an increase in fines may decrease or increase both, or increase one and decrease the other, depending on the mix under consideration.

It follows that the fullest application to control requires preliminary investigations on the particular mixes of interest, but in fact considerable progress is possible without them. This will be illustrated by practical examples in all of which measures of the Bingham constants have been obtained by using the two-point workability apparatus (1,2) to determine the flow curve described by the equation

271

$$T = g + hN \qquad\qquad (1)$$

where T is the torque at an impeller speed N, g is the
intercept on the torque axis and is a measure of yield
value, and h is the reciprocal slope of the line and is a
measure of plastic viscosity.

2   Site investigations without preliminary work

Example 1. Figure 1 shows results obtained on 11
nominally identical batches of a superplasticised flowing
concrete that was being used to cast a basement floor,
and it can be seen that the material supplied was very
variable. The lines marked A form a fan-shaped set
typical of the type obtained when only water content is
varied. Line B, further to the left, can also be

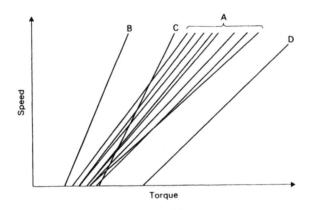

Fig.1. Site results of Example 1.

regarded as belonging to this set, and the simplest
explanation of the high workability of this batch is that
far too much water was added. Line D is quite accurately
parallel to one of the lines in the set and the
difference in this case can be attributed to a failure to
add the correct amount of plasticiser. Line C crosses
several others and this was thought to be due to a change
in the fines content or in the nature of the sand. These
observations were communicated to the project manager who
had obtained the results. He replied that the batch
represented by line D had in fact contained less
superplasticiser (by deliberate decision) and confirmed
that during the making of these batches there had been a
new delivery of sand.
    An alternative way of presenting these results, as a
plot of g against h, is shown in Figure 2. The points
for batches A and B fall close to a single line of

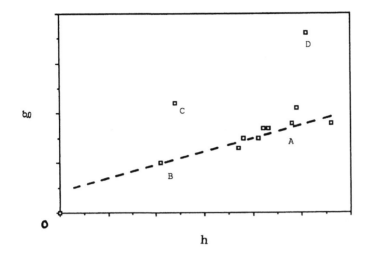

Fig.2. Replot of data from Fig.1.

positive slope, indicating quite clearly that, for these
batches, the only factor contributing to variation in
workability is variation in water content. The points
for batches C and D fall well away from the line showing
that some factor other than water variation is
responsible. This also illustrates a simple technique
that can be used generally. If the relation between g
and h is found to be a simple line with positive slope,
the factor causing variation is water content variation
only. (Theoretically complex combinations of other
factors could result in such a simple relationship but
that is so improbable that it can be ignored). If the
relationship is of any other form, or if there is no
relationship, simple variation of water content can be
ruled out as the cause of workability variation.

Example 2. Results were obtained on 54 batches
delivered to another site, during one day, and the
following deductions could be made.
i. Batch to batch variability was very high
ii. There was some tendency to parallelism of the
flow curves suggesting that variability was due to
variability of plasticiser.
iii. During the day variability decreased. The
coefficient of variation of g was 46% in the morning and
28% in the afternoon. This suggests that either there
was a change in the batcherman or that the batcherman was
gradually learning how to deal with the particular mix
specification.
iv. Changes in workability showed no relation to the
age of concrete at test, although that age varied from
about 15 mins to 75 mins.

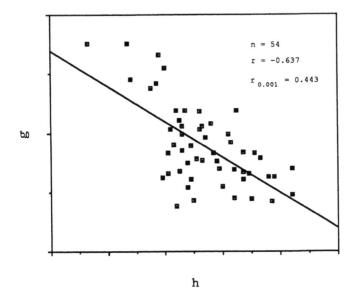

Fig.3. Example 2. Plot of g against h.

v. There was no discontinuous change in workability
that could be associated with some discontinuous change
at the plant, such as delivery of a new batch of
aggregate.

vi. A plot of g against h, shown in Figure 3, showed
a highly significant correlation between the two, but
that correlation was negative, which immediately rules
out variation of water content as the cause of variation
in workability. The reasons for a negative correlation
have not yet been fully elucidated but it is almost
certainly caused by changes in the effectiveness of a
plasticiser; this is referred to again in the next
example.

Example 3. On another site where, as in Example 2, a
lignosulphonate plasticiser was being used, variability
was again thought to be due to the plasticiser because
the flow curves obtained tended to parallelism. Again, a
statistically significant correlation was obtained
between g and h, immediately ruling out water variation
as the culprit, but variation in plasticiser quantity or
method of addition must also be discounted because
measures had been taken to ensure that both of these were
as constant as possible.

Any explanation of a negative correlation must clearly
be in terms of the variation of some factor that affects
g and b in opposite directions, and it was tentatively
suggested that this might be found in a conflict between
the air entraining and plasticising effects of the

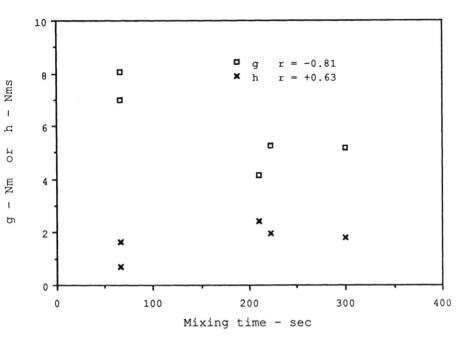

Fig.4. Example 3. Effect of time of mixing.

lignosulphonate admixture used. If that suggestion has
any validity it would be expected that time of mixing
would be important. On one afternoon it was found that
mixing time of 13 batches varied from 1 to 5 mins with a
mean of 3½ mins, and results on five of these batches
were as shown in Figure 4. There was a tendency for g to
decrease with mixing time and h to increase, and although
these relationships were not significant statistically
because of the small number of results, there was nothing
to be lost by recommending standardisation of mixing
time. This was therefore done and seems to have been
justified by the fact that subsequently the ranges of g
and h were reduced by 34% and 15% respectively.

Example 4. The results shown in Figure 5 were obtained
on two separate days on another site, using the apparatus
in its planetary form, and the figures given against the
lines are the final figures of the batch numbers, that
is, they show the order of delivery. Although there are
only about half a dozen results from each day the
following conclusions can be drawn with confidence.
     i. The patterns of results from the two days are
markedly different and sufficiently so as to indicate
some definite change in practice, such as might arise
from the employment of different batchermen.
     ii. Results from the second day form the typical fan-

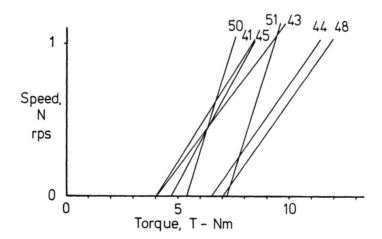

Fig.5a. Example 4. First day.

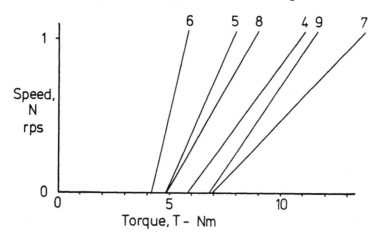

Fig.5b. Example 4. Second day.

shaped pattern for variation of water content only, and
this is confirmed by the fact that the correlation
coefficient between g and h is 0.90 which, even for this
small number of results, is statistically significant.

    iii. The first five of the seven flow curves of the
first day are reasonably parallel to each other, and so
are the last two. This indicates the cause of
variability as being plasticiser variability.

    iv. The last two lines, while being parallel to each
other, cross the other five, which, as already said are
parallel to each other too. This sudden change, between
batches 48 and 50, suggests that some sudden change took
place at the plant. That change could have been, for
example, delivery of a new batch of aggregate.

3   Site investigations with Preliminary Work.

Example 5.   The mix being used for a diaphragm walling
job was specified as 400 kg/cu.m SRPC with 4.5:1 A/C, 41%
fines, and 0.51 W/C ratio.   Before visiting the site
laboratory tests were carried out on the specified mix,

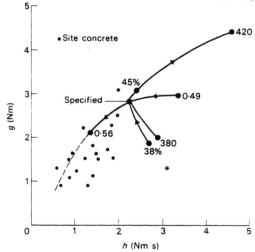

Fig.6. Example 6. Laboratory and site results.

using materials that were to be used on site, and also on
mixes containing more and less water, (0.49 and 0.56
W/C), more and less fines (38 and 45%) and more and less
cement (380 and 420 kg/cu.m), with the results shown in
Figure 6.   This figure also shows results obtained on
site on 20 deliveries of concrete and it can be seen that
nearly all of them are more workable than the specified
mix as shown by the fact that both g and h are less than
they should be, some of them very much so.   In most
cases, for which the points lie close to the line showing
effect of change of water content, or to that line
extrapolated, the higher workability is due to high
water content.   In the case of the batch marked with an
asterisk, the change in workability is probably due to
too low a cement content or percentage fines and,
although it is not possible without further evidence to
say which, it would be easy to decide by means of a
simple sieve test, in a matter of minutes.

4   Establishing of Suitability Bands for Concrete

If it is supposed, as is reasonable, that associated with
any practical concreting process there is some (unknown)
effective average shear rate equivalent to a speed n rps
in the two-point test apparatus, it follows that
concretes suitable for that process will all have the

same apparent viscosity that can be expressed as

$$k = T/n = g/n + h \pm e \qquad (2)$$

where e represents a tolerance. This equation represents a band of suitability on a plot of g against h. If suitability in the process is judged either by some objective measurement or, as is more likely to be possible, subjectively on a numerical scale, plotting of the appropriate points on a g against h plot permits the suitability band to be found, and estimates of k and n to be made.

The most comprehensive study of this type was carried out by Kay (3) on concretes to be considered for use in the hydraulic pressing of slabs. He investigated trial mixes at three levels of A/C ratio, three levels of percentage fines, and five levels of W/C ratio, that is, 45 trial mixes, and also obtained results on production mixes. Two assessments of suitability were made, one on a too wet to too dry scale as judged by an experienced press operator, and the other on a scale of judgement of the finished product in terms of edge and surface defects. The result was the diagram shown in Figure which illustrates quite clearly the suitability band for the process. Further, Kay was also able to estimate that n = 1.94 rps and k = 1.7 $\pm$0.3 Nms. These figures were

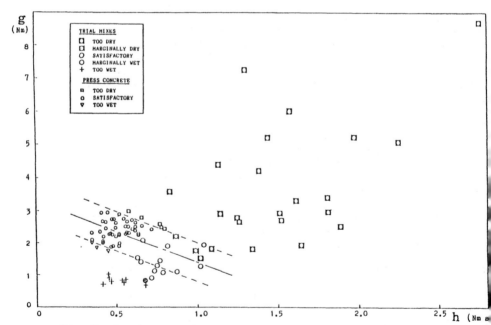

Fig.7. Kay's results, showing suitability band for hydraulic pressing process.

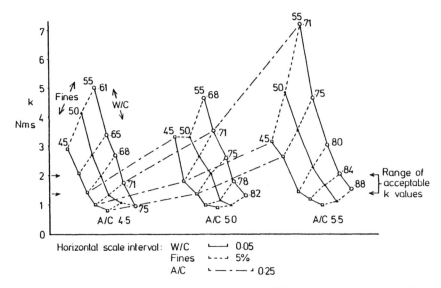

Fig.8. Apparent viscosity at the effective process shear
rate, (calculated from Kay's results) as a function of
mix composition. Showing range of suitable mixes for
hydraulic pressing process.

actually obtained by a consideration of normal production
mixes but have been applied as in Figure 8 to the data
from the 45 trial mixes. This figure is a pseudo-three
dimensional plot which shows the dependence of apparent
viscosity at the process shear rate on each of W/C ratio,
fines content and A/C ratio, while the other two are
constant at various levels, and also shows how those
dependences change as the other two factors change.
    Exactly the same technique could be used for any other
site or works process in which fresh concrete is used,
and could form the basis of an effective control system.

5   Conclusions

Assessment of workability in terms of the Bingham
parameters, yield value and plastic viscosity, can
provide a powerful tool for control of concrete
production. Not only can it fill the role of
satisfactory pass/fail test, which the present Standard
methods are incapable of, but it can also indicate what
are the causative factors of a failure to achieve the
required workability. Moreover, because these factors
may be important in determining the properties of the
hardened concrete, it may make possible early remedial
action to avoid very expensive mistakes.
    Even without any preliminary investigations it is easy
to establish control charts on site so that the purchaser

of concrete can decide whether the material supplied
meets the specification, not only in terms of workability
but also in terms of composition; with results of
preliminary work, the supplier of ready mixed concrete
could go further than this in that he could obtain
information to guide him in taking remedial action for
any deficiency.
   At present, as stated by Barber (4), control of ready
mixed concrete in the UK depends on visual examination
and subjective judgement of the fresh concrete.
Application of the principles outlined here could provide
a means of objective assessment that would be likely to
reduce the incidence of mistakes and consequent disputes.
In principle there would be no difficulty in
computerising the whole process so that values of g and h
obtained automatically would be compared with specified
values and then, on the basis of stored information, any
necessary corrective action would be performed.

6   References

1)   Tattersall, G.H. and Bloomer, S.J. Further
     development of the two-point test for workability
     and extension of its range. Magazine of Concrete
     Research $\underline{31}$ (109) 202-210, 1979.
2)   Tattersall, G.H. and Banfill, P.F.G. Rheology of
     Fresh Concrete, Pitman, London, 1983.
3)   Kay, D.A. The Workability of Hydraulically Pressed
     Concrete. Project Report. Advanced Concrete
     Technology Course. Cement and Concrete Association
     1987.
4)   Barber, P. Quality Assurance - A view from the
     inside. Concrete $\underline{23}$ (11), 24-25, Dec.1989.

Note: An account of the investigations by many workers on
     the factors affecting workability can be found in
     Ref. (2) or in Tattersall, G.H. Workability and
     Quality Control of Concrete, E. & F.N. Spon.
     London. To be published 1990.

29 THE ASSESSMENT OF MIX STABILITY USING THE TWO-POINT TEST

C. ELLIS
Sheffield City Polytechnic, Sheffield, UK
D.E. WIMPENNY
Pioneer Concrete (UK) Ltd, Warrington, UK

Abstract
This paper describes an investigation into the assessment
of the stability of fresh concrete mixes.subjective
assessments of bleeding and cohesion on fresh concrete and
ultrasonic pulse velocity(UPV) measurements on hardened
concrete are related to torque changes in the Tattersall
two-point workability apparatus.in addition, consecutive
changes in 'h' on the same and different concrete samples
are also compared with UPV profiles on short columns.
    There is significant correlation and agreement between
the subjective and objective methods used here to assess
the stability and uniformity of concrete in the fresh and
hardened state.the efficacy of the two-point test for
differentiating between mixes of ostensibly similar
workability is also noted.Further work is required to
clearly identify the changes which occur within the mixes
under test.
Keywords:stability,bleeding,segregation,cohesion,
workability,two-point test,ultrasonic pulse velocity.

1 Introduction and Background

The propensity of concrete mixes to bleed or segregate
during transportation, placing and compaction is of
particular interest in construction.Examples of mix
stability problems range from bleeding at concrete surfaces
and grout leakage from forms to the segregation of piling
mixes during placing.
    Mindess and Young(1981) state that workability is often
defined as the amount of mechanical work or energy required
to produce full compaction of the concrete without
segregation and further refer to bleeding as a special form
of segregation which is essentially the separation of water
from the rest of the mix.some factors which increase
particle segregation also increase bleeding eg mixes which
are too wet or suffer a deficiency in certain particle size
fractions.this subject has been more fully dealt with by
Popovics(1973).
    Both lack of compaction and mix instability may result
in considerable loss in concrete quality measured both in
terms of strength and lack of homogeneity throughout a

pour.The assessment of workability and in particular mix
stability are done frequently on a subjective basis on site,
although methods have been devised to assess stability
objectively as described by Tattersall(1976), Browne and
Bamforth(1977) and Kagaya and Kawakami(1986).Furthermore,
an objective assessment of fresh concrete stability
may also provide an extra test dimension with which to
detect batching errors between mixes with ostensibly
similar workabilities based on British Standards
Institution(1983) methods.

1.1 Applications of the two-point test
The theory and practical applications of the two-point test
developed by Tattersall and its principal advantages over
single point tests have been well documented elsewhere by
Tattersall and Banfill(1983) and Ellis(1984).
    An investigation by Wimpenny and Ellis(1987) showed that
a pressure transducer system attached to the two-point
apparatus and linked to a microcomputer could objectively
remove distortions and reduce operator influence
inherent in the 'manual' method of flow curve (Speed versus
Torque) measurement and the determination of 'g'(yield
value) and 'h'(proportional to plastic viscosity).A
suggestion was made that the change in shape of the
pressure traces might be used to assess the proneness of a
mix to segregation and bleeding.

1.2 Mix stability and the two-point method
It was noted during routine testing of concrete in the
two-point apparatus that some mixes exhibited a change in
shear resistance during consecutive measurements on the
same sample at a similar speed.The opportunity was taken to
monitor this behaviour during the course of a large
experimental investigation, reported elsewhere by Ellis and
Wimpenny(1989), and attempt to relate it to subjectively
and objectively measured characteristics and properties
relating to concrete mix stability and uniformity in the
fresh and hardened state respectively.
    In addition, further work has shown the potential
for using the change in 'h' from consecutive two-point test
flow curves on the same sample as a comparative assessment
of mix stability.

2 Experimental objectives and programme

This work will be presented sequentially as follows:
Section 3 investigates differences or changes between
initial and final torque and regression lines at a single
speed and its relationship with subjective judgment and
ultrasonic pulse velocity (UPV) transit time measurements
for a range of mix designs.
Section 4 investigates the influence of time after water
addition upon the workability of high and low cement

content mixes with similar total 'fines' content (cement plus fine aggregate) by repeated flow curve measurements upon the same samples and upon consecutive samples from the same mix. The changes in 'h' are compared with ultrasonic pulse velocity measurements on short columns as a measure of mix uniformity or homogeneity.

British Standard Institution(1983) manufacturing, curing and testing methods were used throughout the investigations except where otherwise indicated.

3 Relationship between torque change and subjective-objective assessments

3.1 Mix details
This investigation encompassed fifteeen mix designs, including low and high cement contents and slag replacement levels with replication.

Constituents and proportions used in the investigation are described below and in Table 1:

Water:Yorkshire Water Authority
Ordinary Portland cement:Castle cement(Ketton)
Ground granulated blastfurnace slag(GGBS):Civil and Marine
  Ltd. and Frodingham Cement Co. Ltd.
Fine aggregate(FA):Tarmac M sand(Newark)
Coarse aggregate(CA):Tarmac 20-5mm flood plain
  gravel,Blaxton,South Yorkshire.

Table 1 : Concrete mix proportions (kg/m3)

| | Cement content | | | | | | | | |
| | 200kg/m3 | | | 300kg/m3 | | | 400kg/m3 | | |
GGBS(%)	0	40	70	0	40	70	0	40	70
Water	165	165	165	165	165	165	165	165	165
OPC	200	120	60	300	180	90	400	240	120
GGBS	0	80	140	0	120	210	0	160	280
FA	835	835	835	755	755	755	670	670	670
CA	1160	1160	1160	1160	1160	1160	1160	1160	1160

3.2 Testing and results
Torque measurements, using the Two-point test, were made initially at 9 dial speeds; 6,5,4,3.5,3,2.5,2,1.5,1(1.20 to 0.20 rps impeller speed) and then repeated at speed 6.A regression line was fitted through the sequence of decreasing speeds corresponding to torque.The change in resistance to shearing or torque change(Tc) was measured in three ways on the same sample:

TC1- Difference between initial torque and repeated torque measurement at speed 6.
TC2- Difference between torque at speed 6 on regression line and repeated torque measurement at speed 6.
TC3- Difference between initial torque at speed 6 and that on regression line at the same speed.

Subjective assessments (Quantified on a scale 0-10 from low to high) of Cohesion and Bleeding were made together with an objective assessment using UPV transit time difference (range) measurements on 500x100x100mm concrete beams at depths of 25mm to 75mm; measured in the direction of casting.These quantified assessments were plotted against Torque change or difference(Tc); measured in the three different ways.Methods TC1 and TC2 yielded the most significant results.The assessments are plotted against torque change TC1 in Figures 1-3 below.

3.3 Summary findings and discussion
These are listed as follows:

i) The bleeding mark increases approximately linearly with torque change.
ii) In general,the cohesion mark varies inversely with torque change.
iii) Transit time range increases approximately linearly with torque change.

The correlation coefficients for both i) and iii) are significant at the 0.001 level although ii) is significant at the 0.01 level.A linear relationship was assumed in each case.

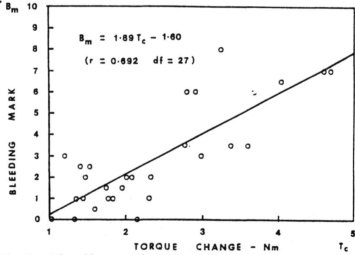

Fig.1. Bleeding mark versus Torque change.

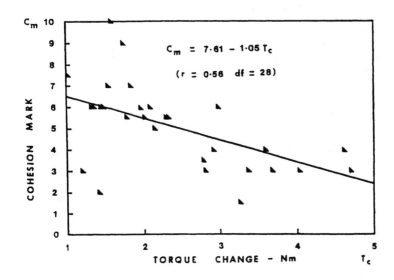

Fig.2. Cohesion mark versus Torque change.

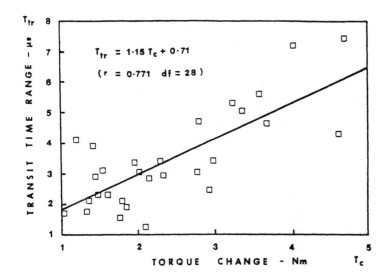

Fig.3. Transit time range versus Torque change.

4 The influence of OPC/FA fraction and time upon flow curve properties and mix uniformity

4.1 Mix details
The mix constituents and proportions used are as described in section 3.1 and in Table 2 but without GGBS. Each mix was mixed for two minutes; due allowance being made for water in the aggregates.

Table 2. Mix proportions and ratios

Constituents	Proportions(kg/m3) Mix A	Mix B		Ratios Mix A	Mix B
Water	175	180	W/C	0.56	0.80
OPC	310	225	A/C	6.25	8.84
FA	770	840	%FA	39.7	42.1
CA	1170	1150			
Totals	2420	2395			

4.2 Testing and results

4.2.1 Fresh concrete
Repeat workability tests were carried out upon two samples taken consecutively from each mix at different times after addition of the water to the mix. Test details and results are included in Figures 4 and 5 and Table 3 for each mix.

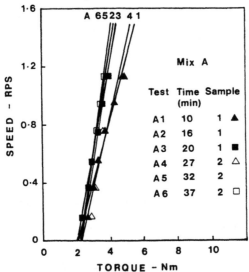

Fig.4. Speed versus Torque curves (Two-point test) - Mix A.

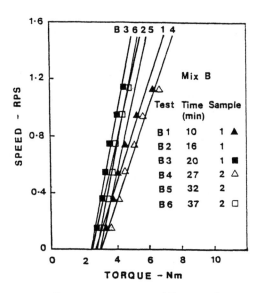

Fig.5. Speed versus Torque curves (Two-point test)- Mix B.

Table 3. Workability tests and results

Test Method	Time after water addition in minutes(Code)					
	10 (A1)	16 (A2)	20 (A3)	32* (A4)	37* (A5)	42* (A6)
				Mix A		
Two-point**:						
g	2.06	2.24	2.09	2.29	2.22	2.25
h	2.32	1.44	1.65	1.99	1.37	1.25
r	0.985	0.991	0.992	0.968	0.991	0.997
Slump(mm):	100	120	-	-	95	-
Compacting Factor:	0.969	-	-	-	0.964	-
				Mix B		
Two-point**:						
g	1.68	1.50	1.41	1.99	1.97	1.80
h	2.80	2.00	1.68	2.94	1.93	1.68
r	0.974	0.994	0.989	0.984	0.992	0.992
Slump(mm):	135	100	-	-	80	-
Compacting Factor:	0.971	-	-	-	0.963	-

NB:- * Fresh sample after remixing for 1 minute.** Based upon six observations for each flow curve using the Mk III apparatus described by Tattersall and Banfill(1983).

4.2.2 Hardened concrete

Manufactured specimens included 2 No. 100mm side cubes for test at each age and one 150x150x450mm column per mix (fully compacted on a vibrating table during two minutes). UPV tests and compressive strength tests were carried out on the cubes at 7 and 28 days.UPV tests on columns were used to assess concrete homogeneity; similar to applications reported by Tomsett(1980).Six horizontal measurements were made at 50mm vertical intervals at each age (three in each direction).Cubes were cured in two conditions;half of them in accordance with British Standard Institution(1983) and the remainder at 60% relative humidity(RH) (figures in parenthesis in Table 4) at 20 C.Columns were stored at 20 C and 60% RH.Results for cubes are given in Table 4 and those for the columns in Figure 6.

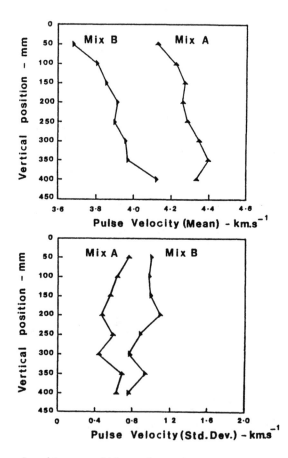

Fig.6. Pulse velocity profiles for short columns.

Table 4. Hardened concrete test results

Test	7day	28day	7day	28day
	Mix A		Mix B	
Density (kg/m3)	2,430 (2,395)	2,440 (2,390)	2,405 (2,345)	2,405 (2,365)
Cube strength (MPa)	31.2 (20.8)	42.9 (26.6)	14.6 (11.2)	23.2 (13.4)
Pulse velocity (km/s)	4.619 (4.269)	4.717 (4.396)	4.430 (3.996)	4.598 (4.202)

4.3 Significance test comparisons

The 't' statistics for changes in 'h' are based upon standard error estimates for 'h' ,described by Tattersall and Banfill(1983),from the flow curve parameters for fresh concrete and are contained in Table 5.Those for pulse velocity(UPV) comparisons at different depths within the columns of hardened concrete, are included in Table 6.The basis of the statistical analysis is after Davies and Goldsmith(1972).

Table 5. Two-point test,Comparison of ' h' values

't' estimates(Significance level)

Consecutive tests/samples

	h A1/A2	h A2/A3	h * A4/A5	h * A5/A6
Mix A	3.91 (0.5%)	-1.46 (10%)	2.26 (5%)	1.15 (>10%)
Mix B	2.33 (2.5%)	1.91 (5%)	3.45 (0.5%)	1.54 (>10%)

Degrees of freedom = 2 x 6 - 4 = 8

* sample after remix for one minute.

Table 6. Comparison of mean pulse velocities at various
vertical positions(depths) within columns

Depth (mm)	Column A		Column B	
	't'	Sig.level	't'	Sig.level
50/400	-5.11	0.1%	-8.80	0.1%
50/100	-2.46	2.5%	-2.33	2.5%
350/400	+1.91	5.0%	-3.18	0.5%

Degrees of freedom = 2 x 6 - 2 = 10

4.4 Summary findings and discussion
These are described as follows:

i) Slump and compacting factor test results for Mix A
and Mix B suggest that there is no significant
difference between the mix properties.
ii) The two-point test flow curve range is larger for
Mix B than for Mix A, suggesting that Mix B may be less
stable than Mix A.
iii) Highly significant decreases in 'h'were recorded
between the first test and first repeat test for each
sample for both Mix A and B but not between first and
second repeats.
iv) The pulse velocity profiles and statistics suggest
lack of homogeneity within both mixes indicating
possible bleeding and or segregation.This trend is most
marked in Mix B.

Examination of the two-point test sample after test
suggested water migration towards the top of the sample
rather than particle segregation as a probable cause of
change in 'h',although the evidence was not conclusive.

5 Conclusions

The preliminary findings from the investigations suggest
that there is significant correlation and agreement between
subjective and objective methods of assessment of mix
stability and uniformity for both fresh and hardened
concrete.Furthermore, the two-point test and the techniques
described provide an additional method for differentiating
between mixes of ostensibly similar workability as well as
for assessing their stability.
    Finally, there is a need to investigate further the
changes which occur within mixes under test in order that
the causes of mix instability might be more precisely
defined.

6 References

British Standards Institution.(1983) Testing Concrete.BS
  1881:Parts 102,103,108,111,116 and 203
Browne, R.D.and Bamforth, P.B.(1977) Tests to establish
  concrete pumpability.j. American Conc. Inst.,Proc.74,
  193-207.
Davies, O.L. and Goldsmith, P.L.(1972) Statistical methods
  in research and production.Oliver and Boyd.Edinburgh.,
  72-195.
Ellis, C.(1984) Relationships between British Standard
  and two-point workability test parameters for OPC/PFA
  concretes.Ashtec '89,Int. Conf. on Ash Technology and
  Marketing 16th-21st Sept.The Barbican,London.
Ellis, C. and Wimpenny, D.E.(1989) A factorial approach to
  the investigation of concretes containing Portland
  blastfurnace slag cements.Third CANMET Int. Conf. on
  Fly Ash,Silica Fume,Slag and Natural Pozzolans in
  Concrete.Supplemetary Papers.19th-24 June.756-775
  Trondheim.Norway.
Kagaya Makoto,Kawakami Makoto(1986) Experimental
  consideration on segregation tendency of concrete
  aggregate particles due to vibration compaction
  (Japanese).J. Society of Materials Science,
  Japan,Vol.35,No.397,1195-1201.
Mindess, S. and Young, J.F.(1981) Concrete. Prentice
  Hall,Inc.,Englewood Cliffs,New Jersey, 203,211.
Popovics, S.(1973) Segregation and bleeding of fresh
  concrete:important properties and their
  measurement.Proc. of a RILEM Seminar held on
  22nd - 24th March,Leeds.The University.Vol.3, 6.1-1 to
  6.1-36.
Tattersall, G.H.(1976)The workability of fresh
  concrete.Viewpoint Publications.London.pp 76-82.
Tattersall, G.H. and Banfill, P.B.(1983) The rheology of
  fresh concrete.Pitman.London.
Tomsett, H.N.(1980) The practical use of ultrasonic pulse
  velocity measurements in the assesment of concrete
  quality.Magazine of Concrete Research.,
  Vol.32:No.110, 7-16.
Wimpenny, D.E. and Ellis, C.(1987) Oil-pressure measurement
  in the two-point workability apparatus.Magazine of
  Concrete Research.Vol.,39:No.140,169-174.

Acknowledgments

The authors wish to thank Frodingham Cement co. Ltd.,Civil
and Marine Ltd. and Sheffield City Polytechnic (in
particular Mr J Proctor, Mr R Hankin and Mr D Gray) in
recognition of their contribution and assistance in
providing the results for this paper.

30 THE EVOLUTION OF FRESH CONCRETE PRESSURE ON FORMWORK WALLS

P.N. OKOH, L. OULDHAMMOU and Ph. BAUDEAU
Institut Universitaire de Technologie, Department of Civil
Engineering, Strasbourg, France

Abstract

The study of concrete pressure, during its placement in a formwork, is approached by considering this heterogeneous mixture as a two-phase medium (cement paste & aggregates). The influence of each of the two phases on the behavior of fresh concrete, with regard to its pressure on formwork as well as the deformation of the formwork wall resulting from this pressure, is considered. This pressure is analyzed with the help of a device, and an experimental method which consists of filling the formwork with a concrete mixture, the composition of which was defined in terms of volume ratio. Results brought to the fore, the following :

- So long as the aggregate concentration in the concrete mix does not attain a certain level, the lateral pressure envelope remains hydrostatic. As from this level onward, the shape of the lateral pressure envelope becomes bilinear : hydrostatic from the free surface to a maximum value, then decreases.

- If the volume of cement paste dominates the concrete mixture, aggregates intervene in the pressure process to a very limited extend.

- Higher lateral pressures are obtained with discontinously-graded concretes in which there are big, coarse aggregates.

- The deformation of formwork wall is essentially due to the mass of concrete contained by the formwork.

- Pressure-deformation diagrams show a linear evolution until a certain range of aggregates concentration in fresh concrete, then tend to stabilize.

Keywords : Fresh concrete, cement paste, aggregates, lateral pressure, deformation, formwork.

1 INTRODUCTION

The study of fresh concrete pressure on formwork walls dates probably back to the last decade of the 19th century. Indeed, one of the first reported pressure studies was in 1894 when E. McCullough <3> measured pressures on a column form. He did this by placing a board on the side of the form and poured concrete into the form until the board broke. Since then, many investigations have been carried out and have enabled a better understanding of fresh concrete pressures on formworks, such investigations include those carried out by P. Baudeau <1> and <2>, R.H. Olsen <4>, N.J. Gardner <9> and the American Concrete Institute <5>.

Formwork is often a temporary structure which has three main functions : give concrete its shape, produce the required surface texture, and support the concrete until it is self-supporting. Moreover, the formwork and its supports must be capable of resisting the forces to which they will be subjected, before and during concreting and during the hardening of the concrete. According to the European Concrete Committee <6>, these forces include : the weight of the concrete, applied loads, climatic actions and the pressure of fresh concrete.

Concrete is a complex structural material, due to its heterogeneous composition and its rheological behavior between the time of mixing and the time of setting. In order to minimize this heterogeneity, we placed our choice on the study of a two-phase concrete composed of :

- a cement paste possessing a rheological behavior that is exclusively viscous,

- a granular phase which resists to shear by a process of aggregate friction.

The two phases enumerated above have precise roles in the mechanical behavior of fresh concrete <7>. The subject of this present study is based on a non-empirical research of the pressure of fresh concrete with respect to the roles played by each of the two phases.

2 EXPERIMENTAL DEVICE AND METHOD

2.1 Measure equipments

The concrete was placed in a steel formwork of 2 m high, 1.35 m large and 0.20 m thick. U2A force captors were arranged on one of the walls of the formwork (thickness = 7 mm), as described on figure 1. These captors are high-precision elements having an error of linearity that is less than ± 0.2 % and a nominal charge of 200 daN. Each of the captors is lodged in an inoxydable steel enclosure that is composed of two elements : a cylindrical element fixed on the formwork wall and which contains a sliding cylinder, and another cylindrical element which contains the captor. The second element is screwed on the first.

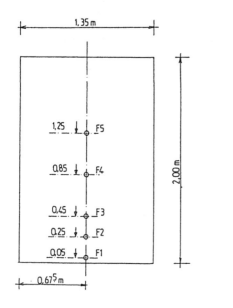

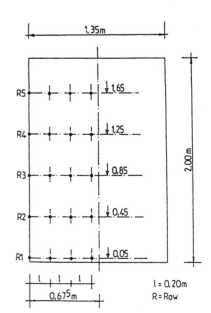

Fig.1 . Position of force   Fig.2 . Position of displacement
       captors .   captors .

The pressure of the fresh concrete is tranmitted to the captor by the help of the sliding cylinder. The sensitive zone is a circle of diameter ø 50 mm.
Twenty W10N displacement captors were installed (see figure 2) in order to measure deformation at various parts of the wall. These captors have a measuring capacity of ± 10 mm and an error of linearity which is less than ± 0.5 %.

Because the formwork is perfectly symmetrical (this was confirmed by preliminary tests with a charge of water), the displacement captors were installed on just a half section of the wall. Also, in order to measure the pressure due to the weight of concrete, a P12 pressure captor was placed at the bottom of the formwork. Its capacity is 0.5 Mpa. A thin, smooth, resistant and flexible sack that perfectly envelops the inner section of the formwork was installed. This enabled to obtain a watertight formwork and to reduce friction between the walls of the formwork and concrete. A UPM 60 measure device was used to scrutinize all pressure and deformation measures.

2.2 Mixtures studied

All concrete mixes were defined by characterizing the two phases cement paste - aggregates by their solid concentration.

- Cement paste :

$$C_c = \frac{V_c}{V_{paste}}, \text{ with :}$$

$$V_{paste} = V_c + V_{ep}$$
$$V_c = \text{volume of cement}$$
$$V_{ep} = \text{volume of water for paste.}$$

It has to be noted that C_c is constant for all mixtures.

- Granular phase :

$$C_g = \frac{V_g}{V_g + V_{paste}}, \text{ where :}$$

$$V_g = \text{volume of aggregates.}$$

This characterization of mixtures by volume ratios became necessary due to the fact that concrete is composed of materials of different densities. A third factor allows to take into consideration the two phases defined above :

$$\nabla = \frac{V_{paste}}{V_g}, \text{ which charaterizes the mix.}$$

2.2.1 Aggregates

It is quite reasonable to think that not all the amount of water used to manufacture concrete goes into the phase of the cement paste. A part of this water also wets the aggregates, and this we have shown using the experimental device on figure 3. The method consists of filtering water, through a constant thickness of aggregates, under the effect of a pressure difference ΔP which is applied between the upper and lower parts of the aggregate sample <7>.

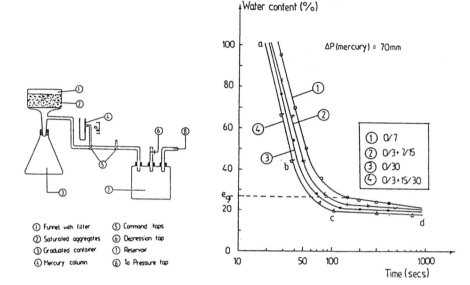

① Funnel with filter	⑤ Command taps
② Saturated aggregates	⑥ Depression tap
③ Graduated container	⑦ Reservoir
④ Mercury column	⑧ To Pressure tap

Fig.3 . Experimental device.   Fig.4 . curves of water filtering
through aggregates .

The quantity of water retained by the aggregates being :

$$Vegr = \frac{\text{Volume of water on aggregates}}{\text{Volume of dry aggregates}} \text{, the following results}$$

were obtained (see fig. 4 and table 1) :

Aggregates	0/7	0/3 + 7/15	0/30	0/3 + 15/30
Designation	G₁	G₂	G₃	G₄
Vegr (%)	26	21	23	11

Table 1 . Quantity of water retained by aggregates.

2.2.2 Cement paste

Knowing that a part of the water used to manufacture fresh
concrete is taken by the aggregates, the amount of water that
really goes into the cement paste (Vep) will be the total
quantity of water used to manufacture concrete (V_E) minus the
amount of water that wets the aggregates (Vegr).

The volume ratio $\dfrac{Vep}{Vc}$ is linked to the definition of cement paste by the following equation :

$$\frac{Vep}{Vc} = \frac{1 - C_c}{C_c} \qquad (1)$$

In the same manner :

$$\frac{V_E}{Vc} = (1 + \frac{Vegr}{100})(\frac{1 - C_c}{C_c}) \qquad (2)$$

2.2.3 The mixture with respect to the theory of T.C. Powers

Powers <10> established a grading of concretes by considering, among others, the water requirement factor that is necessary for concrete to have a given consistency. He chose the following volume ratios as variables :

$$\frac{Vg}{Vc + Vg} \text{ and } \frac{V_E}{Vc + Vg}, \text{ where :}$$

V_E = volume of water
Vc = volume of cement
Vg = volume of aggregates

From these variables, he constructed water-ratio diagrams that enabled him to classify concretes in three categories : AB, BF, FC. For concretes belonging to category AB, the water requirement factor depends mainly on the specific surface area of the constituents. These are rich mixtures. The water requirement factor for concretes belonging to category BF depends not only on the specific surface area of the aggregates, but also on the volume of voids in the mixture. The BF category comprises ordinary mixtures. Finally, for concretes belonging to category FC, Powers stated that their water requirement factor depends only on the voids of the mixture. These are lean concretes.

Figure 5 shows how the various mixtures used in this study are situated, with respect to corresponding diagrams obtained from the water-requirement formulas of T.C. Powers. It should be noted that none of the mixtures studied belongs to the FC category.

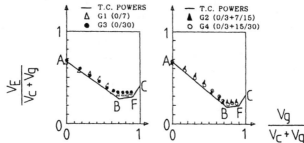

Fig.5 . Mixtures with respect to the theory of T.C. Powers.

2.3 Concreting and experimental method

For technical reasons, the formwork was filled only upto 1.65 m. The aggregates and cement were homogenized for 2 minutes, then mixed with water for 3 minutes. For each aggregate composition cited in table 1, our main objective was to find out the development of the pressure diagram of fresh concrete by a progressive addition of aggregates to the cement paste. The mixture obtained is denoted by the letter M (M1... M9). Cement paste alone corresponds to Mo. The table below shows the corresponding characteristics of each mixture. Three tests were effectuated for each mix M, and only the average results were considered.

Mixture	M0	M1	M2	M3	M4	M5	M6	M7	M8	M9
Cg	0	0.08	0.16	0.24	0.32	0.40	0.48	0.56	0.64	0.71
∇	∞	11.4	5.30	3.15	2.10	1.50	1.08	0.78	0.56	0.42
Density	1.82	1.88	1.94	2.00	2.05	2.10	2.20	2.25	2.30	2.40

Table 2 . Characteristics of each mixture.

3 RESULTS AND ANALYSIS

3.1 Lateral pressure evolution

The evolution of the lateral pressures of concrete, according to the aggregate compositions studied, is shown in figure 6. The lateral pressure diagram for cement paste is hydrostatic. Furthermore, no matter the type of concrete (continuously-graded or not), the lateral pressure diagram remains hydrostatic so long as one does not attain a certain volume ratio of paste-aggregates. Then, the lateral pressure diagram becomes bilinear : hydrostatic from the free surface until a maximum value, and then decreases. The reduction of the pressure slope becomes more pronounced as the aggregate concentration in the mixture increases. It should be noted that the pressure slope reduction is higher in the case of continuously-graded concretes. We also obtained higher pressure values for discontinuously-graded concretes containing big-size aggregates (15/30). This result is in concordance with the one that was previously obtained by Adam, Bennasr and Delgado <8>.

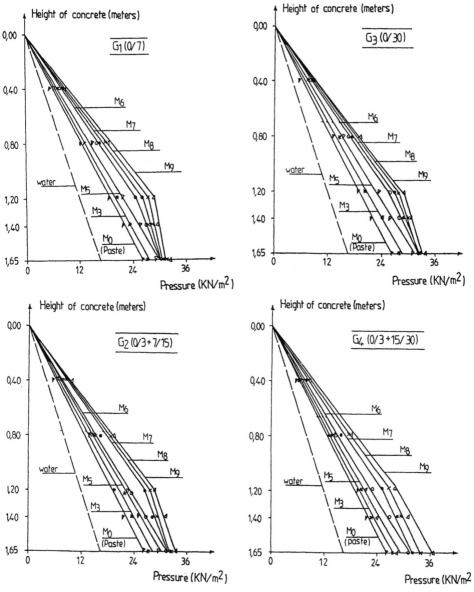

Fig.6 . Evolution of the lateral pressures of fresh concrete
according to its aggregate concentration.

Note : Only 7 of the 10 pressure diagrams are represented, in
order to keep the illustrations clear and easy to read.

3.2 The aggregate effect

By defining the ratio $\emptyset_p$ = Pressure of mixture / Pressure of paste, we wanted to find out if the evolution of the lateral pressure diagrams noticed in paragraph 3.1 could be ascribed to the increase of the granular phase in the various mixtures. Figure 7 $\left\{\emptyset_p = f(\nabla)\right\}$ portrays curves that look alike, each composed of three distinct portions. The ij portion, with a slight difference in $\emptyset_p$ values, signifies that so long as the volume of cement paste dominates the mixture, aggregates intervene in the pressure process in a very limited manner : probably and essentially by their masses. Inversely, the kl portion, possessing a high variation in $\emptyset_p$ values, portrays the presence of an important aggregate concentration in the mixture. The aggregates not only intervene in the pressure process by their masses, but it is also likely that they do so by their spatial arrangements in the mixtures. An intermediary portion jk, situated between the portions earlier described, seems to be a transitional phase in which is realized the passage between the hydrostatic pressure phenomenon (concrete composed mostly of paste) and the bilinear pressure phenomenon (concrete made up in majority of aggregates). It seems therefore that the evolution of the lateral pressure diagram of fresh concrete is directly linked to the volume' ratio between cement paste and aggregates. Our definition of fresh concrete as a two-phase material is thus justified.

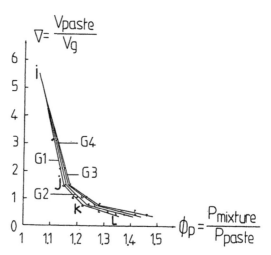

Fig.7 . Influence of the granular phase on the pressure of fresh concrete.

300

3.3 Deformation of formwork walls

At an equal ∇ ratio, practically the same deformation values of the formwork walls were obtained for all concretes. The influence of the mass of concrete on the maximum deformation values of the formwork walls, for all mixtures, is shown on figure 8. The deformation of the formwork wall increases proportionally in function of the mass of concrete. This means, everything being equal, that the deformation of formwork walls depends essentially on the mass of concrete inside the formwork. In the same way as in the case of pressure, $\emptyset_d$ was defined as being the ratio : Deformation due to mixture / Deformation due to paste.

By tracing $\emptyset_p$ with respect to $\emptyset_d$ (fig.9), a pressure-deformation diagram was sought for. It was found that, no matter the type of concrete (continuously-graded or not), the pressure-deformation diagram increases proportionally so long as the mixture presents an aggregate concentration of 60 % or less. Above this value, the pressure-deformation diagram of concrete mixtures tends to stabilize.

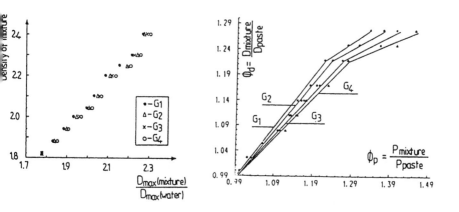

Fig.8 . Influence of the mass of   Fig.9 . Concrete pressure
    concrete on the deformation   versus deformation
    of formwork walls.   of formwork walls.

4 CONCLUSION

By defining fresh concrete as a two-phase medium (cement paste + aggregates), the first series of results enable to show that :

301

- Cement paste exerces a hydrostatic lateral pressure on formwork walls. When aggregates are progressively added to cement paste, hydrostatic pressure is still obtained until the aggregate concentration in the mixture attains a certain limit. Above this limit, the pressure diagram become bilinear ;
- So long as the volume of cement paste dominates that of aggregates, the latter intervenes in the pressure of the concrete mixture in a very limited manner ;
- The deformation of formwork walls is essentially due to the mass of concrete contained by the formwork ;
- For concretes with an aggregate concentration of 60 % or less, lateral pressure evolves linearly with respect to the deformation of formwork walls. Above this value, the pressure-deformation diagram tends to stabilize.

For a further analysis of the above results, notably in what concerns the evolution of the lateral pressure diagram of concrete, an experimental study was realized which enabled to prove that the granular phase provokes the change of pressure diagram. Secondly, two pressure calculations were effectuated using a modelling process which took into account the finite elements method. The results of these studies will be presented shortly.

REFERENCES
⟨1⟩ Baudeau, Ph. and Zygler, M (1976) Les bétons de Parois moulées - Miniaturisation d'un bétonnage, Bulletin de Liaison LCPC, Ref. 1820.
⟨2⟩ Baudeau, Ph. and Weber, J.D. (1980) Comportements du béton frais de fondations à parois moulées dans le sol, Annales I.T.B.T.P. N° 380.
⟨3⟩ McCullough, E. (1908) Reinforced concrete. A manual of practice, Clark Publishers company, U.S.A.
⟨4⟩ Olsen, R.H. (1968) Lateral pressures of concrete of formwork, Ph.d thesis, Oklahoma State University U.S.A.
⟨5⟩ Hurd, M.K. (1979) Formwork for concrete. 4th Edition, ACI Committee 347.
⟨6⟩ C.E.B. (1977) Manuel de technologie - Coffrage N˙ 115.
⟨7⟩ Barrioulet, M. (1977) Contribution à l'étude des rôles respectifs joués par la pâte interstitielle et par les granulats dans le comportement du béton, avant la prise du liant (Ph.d thesis).
⟨8⟩ Adam, M. Bennasr, M. and Delgado, H. (1965) Poussée du béton frais sur les coffrages, Annales I.T.B.T.P. N˙ 207-208.
⟨9⟩ Gardner, N.J. (1984) Formwork pressure and cement replacement by fly ash, Concrete International, pp 50-55.
⟨10⟩ Powers, T.C. (1963) The Properties of fresh concrete, Wiley & Sons Inc., New York.

31 PREDICTION OF CONSISTENCY OF CONCRETE BY USING KEY PROPERTIES OF RAW MATERIALS AND MIX COMPOSITION

V. HANKE
Verein Deutscher Zementwerke, Düsseldorf,
Federal Republic of Germany

Abstract
This paper gives a model to predict the consistency of concrete. Key-properties of raw materials and mix composition are discussed on concrete spread, measured with the flow table according German standard. Influences of aggregate, quartz fillers and cement paste are illustrated. Also it is described, that the workability of fresh concrete depends on the grading and water demand of all components as well as on the water content. Calculating a "free water content" out of this parameters, a good correlation to the practical behaviour is reached.
Keywords: Concrete, Consistency, Cement, Filler, Grading of aggregate, mix composition, workability

1 Introduction

In the fifties of this century, when the well known mix design methods were developed, concrete was produced at the building site with aggregates, cement and water. Today more than 80 % of concrete products are mixed in plants using admixtures like plasticizers and additives such as PFA. Furthermore, aiding the production of concrete by using computers, nowadays it has become much easier to regard the influences of all used components, which are interacting in some cases.
    Produced by using aggregates, admixtures, additives, cement, and water concrete has become one of the most usual building materials. The accurate adjusting of consistency is necessary for placing and compacting fresh concrete properly. In the established mix design methods concerning workability, the water demand of the aggregate is calculated by adding up the experienced and tabulated water demands of the used aggregate fractions. That is just a very rough estimation, because adding up the single water demand values does not take the actual interactions between different fractions into account.
    Summarizing the facts from above, it is necessary to develop a model by using key-properties of raw materials and mix composition to predict the properties of concrete. Regarding the behaviour of fresh concrete, it is important to know that most of the shown parameters are valid for hardened concrete too.

2 Modelling the properties of concrete

Binding aggregate with cement paste, concrete is a material that changes its properties during time. Cause of this especially cement, as the main reactive component, is responsible for the changing behaviour of concrete. Concerning to fresh concrete's workability in early ages, the attributes of cement and inert fillers can be treated nearly in the same way. In spite of that the influence of cement in hardened concrete is completely different.

Regarding the consistency, concrete is often idealized by a model of spheres. Herein all solid components are approximated to spheres surrounded by layers of water. In this way the consistency of those mixtures depends on the mobility of all solid particles as well as on the water content. Closing to the particle size distribution of fines and the grading of aggregate, the storage of spheres has an influence on the mobility of the solids. In a 3 year research project in the Research Centre of the German Cement Industry the influence of filling spaces between the particles by "fitting" diameters was systematic investigated.

3 Research program

To determine the parameters about the consistency of concrete, more than 200 experiments were done. Besides, optimizing the grading of the aggregate, the main work concerned the particle size distribution of fines including binder. Table 1 shows the specific surface measured by Blaine test and water demands of all cements and quartz fillers used in this investigations. In addition, the initial and final setting time of all cements is given as well as their compressive strengths. All values are measured using German Standard DIN 1164. Only the water demand of quartz 2 could not be determined, because of the extreme fineness of this material.

Table 1. Attributes of cement.

Cement/Filler	Specific Surface [cm²/g]	Setting Time Initial [h:min]	Final [h:min]	Water Demand [mass-%]	Compressive Strength [N/mm²]
PZ35F (OPC)	3110	2:20	3:00	26.0	47.9
PZ45F (OPC)	4090	2:00	2:40	30.0	59.4
HOZ35L (BFSC)	3310	3:00	4:00	26.5	48.7
HOZ45L (BFSC)	4000	2:40	3:20	25.5	56.2
Quartz 1	107			27.5	
Quartz 2	1450			-	

In all experiments the consistency of concrete was tested by using the concrete spread on flow table according DIN 1048. The minimum and maximum content of water, cement and quartz filler as well as the ranges of water cement ratio is shown in table 2 . The main influences on the consistency of concrete, using aggregate ,cement, fillers and water, are illustrated in the following paragraphs on characteristic results.

Table 2. Mix compositions .

	Unit	min	max
Water content	l/m^3	180	220
Cement content	kg/m^3	260	440
Quartz filler content	kg/m^3	0	300
Water cement ratio	-	0.50	0.70

4 Results

4.1 Influence of aggregate

Regarding the quantity, aggregate is the main component of concrete. Its content is calculated by the volume that is necessary to fill a cubic metre. Therefore, only the grading of aggregate can be changed, if the contents of water and cement are given. Figure 1 shows the influence of aggregate's grading on concrete spread tested 10 minutes after mixing. For all values the same Ordinary Portland Cement (PZ35F), water cement ratio of 0.63 and cement content of 300 kg/m^3 was used. Varying the grading of aggregate in 4 steps from coarse (A32) to fine (C32) the measured concrete spread falls about 20 cm. Furthermore, the part of fines (diameters less than 0.25 mm) is given in brackets. The increase of fines is closely bound up with decreasing consistency values.

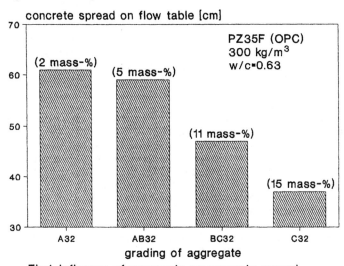

Fig.1. Influence of aggregate on concrete spread.

305

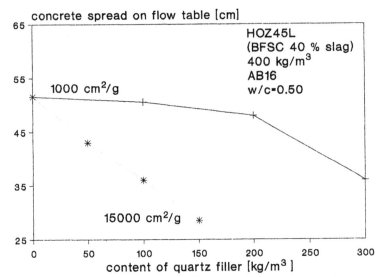

Fig.2. Influence of quartz filler on concrete spread.

4.2 Influence of quartz filler

Analysing the facts from above, the relationship between concrete
spread and the content of quartz filler is illustrated in figure 2.
All mixes were made with blast furnace slag cement, containing 40
mass-% of slag, with a content of 400 kg/m^3. Furthermore the water
cement ratio of 0.50 and the grading of aggregate (AB16) was in all
experiments the same. Two quartz fillers, with specific surfaces of
1000 cm^2/g and 15000 cm^2/g were added in contents ranging from 0 to
300 kg/m^3. It can be recognized, that an addition of quartz filler
always leads to lower concrete spreads. Adding the very fine quartz
2, workability changes to smaller values than using quartz 1 by the
same content.
Just varying the content and specific surface of the added inert materials, the influence of quartz fillers is explained in figure 2 exem-
plary. Additional researches, using other cements of table 1 as well
as changing contents of cement and water, results always in the same
stiffening of consistency, if only inert quartz fillers were added.

4.3 Influence of cement paste

It is generally known that compressive strength, as an important property of hardened concrete, depends on the water cement ratio mainly.
Regarding to workability an improvement can be reached by adding water. In all, rising up the content of cement paste without changing
of water cement ratio will be one possibility to vary workability of
concrete by keeping compressive strength constant. This is the reason, why figure 3 shows the influence of cement paste on concrete
spread. Using always the same materials and water cement ratio, the
mix composition varied by content of cement paste only. Rising the
volume of lime, the consistency of concrete changes from stiff to
fluid. This means in dimensions that an adding of 1 vol-% of paste in

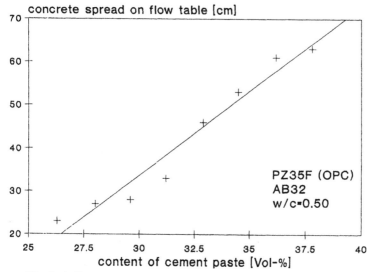

Fig.3. Influence of cement paste on concrete spread.

creases the concrete spread value about 3 cm.

5 Discussion

Regarding concrete as a model of spheres, consistency depends on characteristic attributes of solids and the water content. Considering a mixture with invariable material attributes, a rising water content increases the thickness of water layers around every particle. Cause of this, the solids get higher mobility, resulting in fluid consistency of concrete.

Concerning workability, the water demand including physical and chemical binding of water is a very important attribute of materials. From the physical side water is bound by surface tension and absorptive strength. Using aggregate dried at the surface, the influence of absorptive strength was equal for all experiments and can be neglected here. The surface tension depends on the diameters of particles as well as the specific surface. Therefore, the water demand values are mainly influenced by the particle size distribution of the materials. Furthermore, reactive components, mainly cement, have the property to bind water on chemical base.

In all, the difference between water content and water demand of the mixture seems to be important for the consistency of concrete. Figure 4 shows the relationship between this "free water content" and the tested concrete spreads. Fitting the test values by linear regression a good correlation can be reached. This means that the variation is within the small range of 3 cm at all.

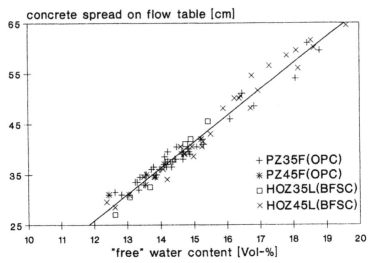

Fig.4. Relationship between "free" water content and concrete spread.

6 Conclusions

As mentioned above the consistency of concrete depends on the physical and chemical behaviour of the raw materials as well as on the mix composition. This means that the concrete spread tested on flow table and the free water content, calculated by the water content and water demand values correlated very well. To predict the consistency of concrete using cement, water and aggregate including inert fillers the following key-properties are necessary:

(a) water content
(b) cement content
(c) content of fines (≤0.25 mm)
(d) grading of aggregate
(e) particle size distribution of fines
(f) reactivity of cement

It is planned to insert the influences of admixtures and plasticizers in this model step by step. Also, knowing that the behaviour of hardened concrete is related to the consistency of concrete, the model will be expanded on further properties of concrete. Regarding to the compressive strength first results using the above mentioned model showed good correlations as well.

7 References

Bonzel, J. and Dahms, J. (1978) Über den Wasseranspruch des Frischbetons (About the water demand of fresh concrete). **Betontechnische Berichte**, 19, 121-156.
Fuller, W.B. and Thompson, S.E. (1963) The Laws of Proportioning con-

crete. **Concrete**, 1053, 67-172.
Kluge, F. (1949) Vorausbestimmung der Wassermenge bei Betonmischungen für bestimmte Betongüten und Frischbetonkonsistenz. (Predicting the water content of concrete mixes regarding to quality and workability). **Der Bauingenieur**, 6, 172-175.
Krell, J. (1985) Die Konsistenz von Zementleim, Mörtel und Beton und ihre zeitliche Veränderung (The constency of cement paste, mortar and concrete regarding to their temporal changing). **Schriftenreihe der Zementindustrie**, 46.
Springenschmid, R.,Ganser, A. and Schrage, I. (1988) Effect of Ultra-Fine Aggregate Particles on the Structure and Durability of concrete. **Massivbau Baustofftechnologie**, 6, 23-30.
Zimbelmann, R. (1987) A method for strengthening the bond between ce ment stone and aggregates. **Cement and Concrete Research**, 4, 651--660.

PART SEVEN

INFLUENCE OF VIBRATION ON CEMENT-BASED SYSTEMS

2 INFLUENCE OF VIBRATIONS ON THE RHEOLOGICAL PROPERTIES OF CEMENT

L. BÖRGESSON and A. FREDRIKSSON
Clay Technology, Ideon Research Centre, Lund, Sweden

Abstract
A new technique for cement grouting of fine rock fractures has been developed at Clay Technology in Lund. The basic ideas of the technique is to use low water cement ratios and to lower the viscosity of the grout by vibrations.
The paper shows test methods and results from viscometer tests where the bob is rotating in a vibrating cup. The tests show e.g. that the stiff non-newtonian cement becomes nearly newtonian when the amplitude is high enough. They also show that low frequencies and high amplitudes are more efficient than vice versa at constant power. The influence of shear rate and vibrating shear strain amplitude on the shear resistance is mathematically formulated in two equations.
The cement is injected by using a high static pressure superposed by a dynamic oscillating pressure. The oscillating pressure is transformed to oscillating displacements when penetrating a fracture. Two mathematical theories of the oscillating compressible flow of the grout in a fracture have been developed, one analytical and one numerical. The theories have been checked by large scale slot injection tests using slots with apertures of 100-300 μm.
The technique has been successfully tested by full scale injections in the Stripa mine.
Keywords: Cement, Rheology, Vibrated cement, Vibrations, Grouting.

1 Introduction

The need for grouting very small fractures with a material that can survive for thousands of years has arisen in connection with the developement of methods for disposal of radioactive and other toxic wastes. Granitic rock surrounding waste packages at a depth of ≈500 meters often contains fractures with apertures smaller than 100 μm that can lead water. If these small fractures can be effeciently sealed the function of the repository will be considerably improved.
Laboratory as well as field tests of two possible grout materials have been performed: cement and smectite-rich

clay. The rheological propeties of these materials are
similar in some ways but differ in other ways. Only the
cement will be treated in this article.

In order to increase the longevity of the cement it is
vital to use a cement with a low water/cement ratio.
However the viscosity of the grout must also be low since
otherwise the cement will not be groutable. The problem of
having low water/cement ratio, low viscosity and deep
fracture penetration at grouting has been solved by adding
superplasticizer to the cement and by using a combination
of high injection pressure and an oscillating pressure
superimposing the static pressure.

2 Cement Composition

The cement used in these tests is a finely ground
injection cement of Portland type. Additives of 0 - 10 %
silica fume and 0 - 2 % superplasticizer were used. The
silica fume is added to increase the chemical stability
and the superplasticizer is added to decrease the
viscosity. The water/cement ratio was varied between 0.3
and 0.5.

3 Rheological testing

The rheological properties were tested in a Brookfield
viscometer with a rotating vertical bob inside a
cylindrical cup. The cup was vibrated vertically by an
electromagnetic vibrator with frequencies f that can be
varied between 20 and 10 000 Hz and displacement
amplitudes δ_a that can be varied between 0 and 1.5 mm. The
amplitude was measured by a small accelerometer fixed to
the cup. Fig 1 shows the arrangement

The width b of the slot between the cup and the bob was
varied in a number of tests. These tests showed that
neither b nor the vibrating amplitude δ_a but the ratio
δ_a/b, which is equal to the amplitude of the vibrating
shear strain γ_a, is the parameter that determines the
influence of the vibrations on the rheological behavior.
The width b was then fixed to 3 mm and the maximum
vibrating shear strain amplitude used in the tests was
$\gamma_a=0.5$.

4 Rheological properties

Different cement compositions with varying water/cement
ratio, superplasticizer content and silica fume content
have been tested at different frequencies, amplitudes and
at different times after mixture. Most tests however have
been performed 30 minutes after mixture since this

Fig. 1. Test arrangements for the rheological tests on vibrated cement. To the left is the sine wave generator. In the middle the viscometer and the vibrator is seen. A small piezotron accelerometer is fixed to the cup. The piezotron coupler and the oscilloscope is seen to the right.

corresponds to the required time at field grouting.

An example of different series of viscometer tests performed under vibrations with the frequency $f=40$ Hz at different amplitudes varying between $0 \leq \gamma_a \leq 0.5$ is shown in Fig 2. The cement contains 10% silica fume and 0.75% superplasticizer and has a water/cement ratio of 0.35.

Fig. 2 shows that the material is not newtonian in the unvibrated state. When vibrations are applied the shear resistance is decreased and the shear-stress/shear-rate relation obviously becomes a straight line in the double logarithmic diagram. The relation can be written according to Eqn 1:

$$\tau = m \cdot (\dot{\gamma}/\dot{\gamma}_0)^n \qquad (1)$$

where τ =shear stress (Pa)
$\dot{\gamma}$ =shear rate (s^{-1})
$\dot{\gamma}_0$ =normalized shear rate (=1.0 s^{-1})
m =parameter (shear stress at $\dot{\gamma}=1.0$ s^{-1})
n =parameter (inclination of $\tau-\dot{\gamma}$ line)

The inclination is $n \approx 0.9$ in the example in Fig 2. Several tests have shown that n varies between 0.9 and 1.0 when the vibrations are large enough to destroy the structure of the material. $n=1.0$ means that the material

is newtonian. The vibrations thus make the material close
to newtonian. Fig 2 also shows that the vibrations lower
the shear resistance: the higher the shear strain
amplitude γ_a is the lower the shear resistance will be.

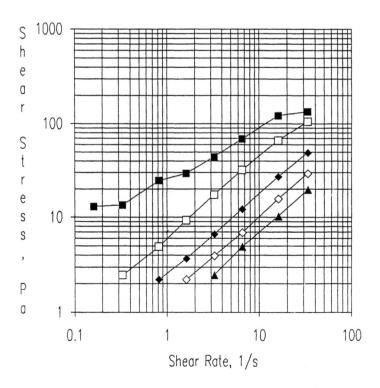

Shear Rate, 1/s

Fig. 2. Shear stress from viscometer tests as a function
of shear rate for different applied vibrating
shear strain amplitudes γ_a. f=40 Hz; w/c=0.35;
0.75% superplasticizer; 10% silica fume.
■ γ_a=0.0, □ γ_a=0.067, ◆ γ_a=0.2, ◇ γ_a=0.33,
▲ γ_a=0.5

Fig 3 shows the parameter m in Eqn 1 plotted as a
function of the applied oscillating shear strain amplitude
γ_a at different applied frequencies f. The m-γ_a relation
is obviously also a straight line in a double logarithmic
scale. The relation can thus be written according to Eqn
2:

$$m=a\cdot (\gamma_a/\gamma_{a_0})^b \qquad (2)$$

where γ_a =oscillating shear strain amplitude
γ_{a_0} =normalized γ_a (=1.0)
a =parameter (Pa)
b =parameter (m at γ_a=1.0)

Fig 3 also shows that a decrease in γ_a can be well compensated by an increase in f up to 300 Hz. At f=1000 Hz however the required amplitude γ_a is as high as at f=300 Hz. An increased frequency as well as an increased oscillating shear strain amplitude are thus decreasing the shear resistance.

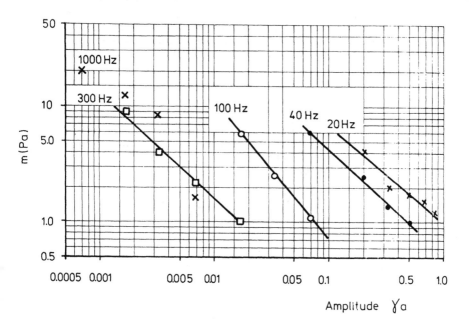

Fig. 3. The parameter *m* plotted as a function of the oscillating shear strain amplitude γ_a at different applied frequencies f. Cement type: see Fig 2.

The effect of an increased amplitude versus an increased frequency can be compared if f is plotted as a function of γ_a at the same rheological properties, that is at the same value of m. Fig 4 shows the γ_a-f relation at m=1.0. The figure shows that 10 times change in amplitude correspond to 5 times change in frequency at f<300-400 Hz. This means that the effect of doubling the frequency is higher than the effect of doubling the amplitude. However the required power is direct proportional to f^3 and γ_a^2 which means that more power is required if the frequency is increased than if the amplitude is increased in order to reach the same decrease in shear resistance. Thus a low frequency is favourable and it should not exceed $\approx$300 Hz to get a god effect on the shear resistance.

5 Fracture penetration

The theories and laboratory results accounted for in chapter 4 show that a vibrating shear strain with an amplitude $\gamma_a > 0.01-0.1$ significantly reduces the shear resistance. Such vibrations can at grouting be produced by superposing a static and a dynamic pressure. The dynamic pressure pulses will propagate from the injection machine via the grout filled bore hole into the fracture into which the grout is penetrating. In the fracture the pressure pulse will to some extent be transformed to a displacement pulse governed by the compressibility of the

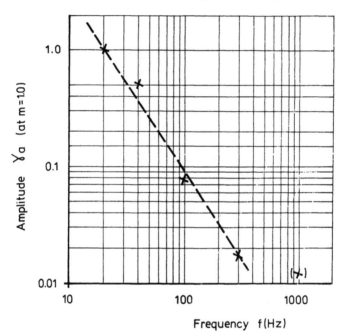

Fig. 4. The vibrating shear strain amplitude as a function of the corresponding frequency at equal viscous behaviour ($m=1.0$)

grout. Fig 5 shows the amplitude of the velocity profile of the pressure pulse in a fracture. Only the oscillating part of the velocity is shown. The grout is thus vibrating backwards and forewards with no displacement close to the rock surface and maximum displacement s_{max} in the centre of the fracture. The vibrations are thus producing an oscillating shear strain amplitude

$$\gamma_a = \tan\alpha = s_{max}/(\mu/2) \qquad (4)$$

where μ is the fracture aperture. Since the fracture apertures are very small ($\mu \approx 0.1$ mm), it means that the

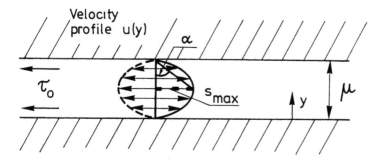

Fig. 5. Velocity profile produced by the oscillating
pressure in a fracture with the aperture μ.

displacement s_{max} only need to be 0.001-0.01 mm in order
to produce a shear strain amplitude large enough to reduce
the shear resistance of the grout.

6 Penetration calculations

Theoretical studies of the dynamic response and the rate
of grout penetration have been conducted in order to
understand the process and predict penetration.

Analylical as well as numerical solutions describing
the inflow and wave propagation in fractures have been
derived assuming a sinusoidal pressure oscillation. The
theories are described by Pusch et al. (1988) and
Börgesson & Jönsson (1990).

The grout flow theory makes it possible to predict the
penetration of grout into a fracture as a function of time
if the fracture geometry is known. However there are some
theoretical problems, the most important one being the
influence of the shear strain amplitude on the viscosity
of the grout. The shear strain amplitude of the grout in
the fracture, caused by the oscillating pressure amplitude
in the beginning of the fracture, is varying with
penetration depth as well as along the fracture. This
means that the viscosity will vary in the same way and the
situation will be very complex. This complication can be
diminished by using a simplified theory in which the
average viscosity along the fracture is used and thus only
the change in viscosity with time (and thus penetration
depth) will be taken into account. A correct solution can
be achieved by using numerical calculations but since
these methods have yielded similar results the simplified
technique is usually applied.

7 Grout tests

The material model and flow theories have been tested in artificial fractures with apertures 100 μm and 300 μm, length 3.0 m and width 5 cm. The walls of the fracture are very stiff (5 cm solid steel) and the surfaces are rough. The fracture is open in one end and connected to the injection device in the other end. Pressure transducers and small windows are applied along the fracture in order to study the induced stresses and the rate of grout penetetration.

The injection device have been especially designed to achieve a static high pressure superimposed by a dynamic oscillating pressure. The oscillations are achieved by percussion. A static pressure of 1-5 MPa and an oscillating pressure with the amplitude 0.1-5 MPa and the frequency 40 Hz were used. Fig 6 shows a picture of the device. A colloid mixture, a stirrer and a pump are supplying the machine with grout.

Fig. 6. The dynamic injection device
A) Colloidal mixer; B) Stirrer; C) Pump; D) Percussion machine; E) Piston; F) Grout cylinder; G) Pneumatic cylinder for static back pressure; H) Injection pipe; J) and K) Valves

Fig 7 shows a comparison between a test in the 100 μm fracture and the corresponding computation of the entrainment. It is obvious that the computed process differs somewhat from the measured one although the final entrainment is of the same order.

The penetration as a function of the fracture aperture

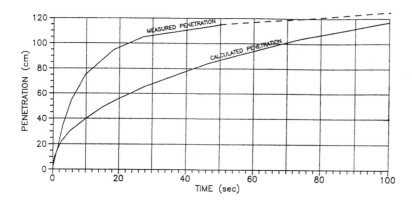

Fig. 7. Comparison between measured and calculated
penetration into a fracture with the aperture 100μm

and as a function of the viscosity of the grout can be
calculated according to theories mentioned. The
penetration depth is mainly a function of the fracture
aperture, the imposed static pressure, the amplitude of
the oscillating pressure and the parameters m and n in Eqn
1. Fig 8 gives a qualitative idea of the penetration as a
function of the fracture aperture for different m values
if $n=1$, $f=40-200$ Hz, the static pressure 2 MPa and the
oscillating pressure amplitude 5 MPa. The grout is
composed to give $m<1.0$ Pa at $\gamma_a=1.0$.

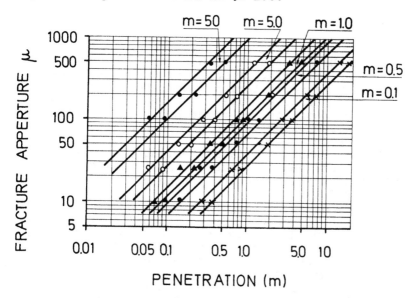

Fig. 8. Influence of the viscosity (m, Pas) and the
fracture aperture on the penetration depth.

8 Field tests

Several full scale field tests have been performed in the crystalline rock at the test site in the Stripa mine. Bore-holes with 76 mm diameter as well as large holes with 0.76 m diameter have been successfully grouted. Permeability tests before and after grouting have shown that the average hydraulic conductivity of the rock can be reduced to 10^{-10} m/s irrespective of the original value. Fractures with apertures as low as 20-30 μm have been injected to several decimeters depth.

9 Conclusions

Vibrations reduce the shear resistance of cement considerably if the right combination of frequency and oscillating shear strain amplitude is applied. The best efficiency is reached at frequencies below 300 Hz. The technique can be applied at grouting if the oscillations are achieved by pressure pulses from percussion under high static pressure. The pressure pulses will, in the fractures, be transformed to displacements with oscillating shear strain amplitudes which are high enough to strongly reduce the shear resistance of the grout.

Large scale tests in artificial fractures as well as field tests have proven that the technique is applicable.

10 References

Börgesson, L. and Fredriksson, A. (1988) Rheological properties of Na-smectite gels used for rock sealing. **Workshop on artificial clay barriers for high level radioactive waste repositories.** Lund, Sweden.

Börgesson, L. and Pusch, R. (1989) Sealing of fractured rock; grout composition and grout properties. **NEA/SKB International symposium on the Stripa project.** Stockholm, Sweden.

Börgesson, L. and Pusch, R. (1989) Sealing of fractured rock; pilot tests and full scale tests. **NEA/SKB International symposium on the Stripa project.** Stockholm, Sweden.

Börgesson, L. and Jönsson, L. (1990) Grouting of fractures using oscillatory pressure. **International conference on mechanics of jointed and faulted rock.** Vienna, Austria.

Jönsson, L. (1989) Computation of high-viscosity transient flow in small fractures. **6th international conference of pressure Surgs.** Cambridge, England.

Pusch, R. et al. (1988) Rock sealing - interim report on the rock sealing project. **Stripa project, technical report 88-11.** SKB, Stockholm, Sweden.

33 EFFECT OF VIBRATION ON THE RHEOLOGICAL PROPERTIES OF FRESH CEMENT PASTES AND CONCRETES

G.H. TATTERSALL
Department of Civil and Structural Engineering,
University of Sheffield, UK

Abstract
This paper summarises and collates already published
work on the use of the two-point workability apparatus
to study the effect of vibration on the flow curves of
cement pastes and fresh concretes. It is shown that,
provided certain threshold conditions are exceeded,
yield value is reduced to zero and the flow curve of
the vibrated material can be described by a power law
pseudoplastic relationship. The important parameter
for assessing the efficiency of vibration is the
maximum velocity.
    At very low shear rates the vibrated material may
be regarded as Newtonian and this leads to the
development of a very simple method of investigation
which has been used over a wide range of frequencies
and amplitudes. Again, the important parameter is
velocity.
    It is shown that there is a good correlation
between results, for Newtonian viscosity, obtained by
two completely different experimental methods.
    Implications for practical design of vibrators are
discussed briefly.

Keywords: Vibration, Two-point test, Workability, Flow
curve.

Introduction

It has long been common practice to use vibration in
the placing and compacting of concrete and the
importance of the topic is illustrated by the amount of
research that has been carried out, and by the regular
appearance of review and summary statements (1-4).
Although some measurements were made of vibration
parameters within the fresh concrete, (5,6) much of the
early work depended on measuring properties of hardened
concrete that had been cast under a variety of
vibration conditions (7,8). Suggestions were made that
the important parameter was the maximum acceleration of
the applied vibration but, for example, Cusens (8) also

found that there was a threshold amplitude below which there was no effect.

Although it may be argued that the most profitable course would be to investigate effects on the fresh concrete, little work was done in this direction, with the exception of some empirical flow tests that had no theoretical basis, and the soil mechanics approach of L'Hermite (9) who used an annular shear box. L'Hermite also developed a theory but that was shown (10) to be defective on both physical and mathematical grounds.

The apparent liquefaction that occurs when vibration is applied to fresh concrete suggests the possibility of the occurrence of a thixotropic change in the cement paste, but experiments with a coaxial cylinders visocometer (10,11,12) showed that although cement paste suffered a structural breakdown under shear, that breakdown was not reversible, so some alternative explanation must be sought.

Nevertheless, just as ordinary observation of unvibrated fresh concrete suggests that the material possesses a yield value (at least in practical engineering terms) so ordinary observation suggests that when vibration is applied that yield value is much reduced, at least to an extent such that the material can flow under its self-weight. The results of L'Hermite referred to above afford support for this qualitative observation, as do the more recent ones of Lasalle and Legrand (13). They measured the minimum torque required to turn a paddle placed at various distances from a vibrating source in a calcite paste (used instead of cement paste to avoid problems arising from setting) and showed that, over a range of distances, yield value was effectively destroyed, but, because of the nature of their apparatus, no information was obtained about torques at higher shear rates. It is also necessary to note that results obtained on calcite pastes may or may not be applicable to cement pastes.

Development of the two-point workability apparatus (14,15) presents the possibility of further advance because it is a relatively simple matter to mount the bowl of apparatus on a vibrating table so that the complete flow curve of a vibrated concrete may be determined and compared with the flow curve of the same concrete in the absence of vibration.

Preliminary Experiments

Preliminary experiments were carried out on a mix with an aggregate/cement ratio of 6 and with 40% fines, in

which all the cement had been replaced by pfa to reduce
the rate of change with time. Water content was such
as to give 150mm slump. The bowl of the workability
apparatus was mounted on a vibrating table driven by a
rotating eccentric formwork vibrator whose amplitude
could be varied by altering the relative angular
position of the eccentric masses. The concrete was
placed in the bowl and the flow curve was obtained in
the normal way when the vibrator was not operating.
The impeller was then run at the top speed used and the
vibrator was switched on, resulting in a very rapid
drop in torque, followed by a slower rise. The rise
was attributed to the onset of compaction, so the
minimum torque observed was taken as measure of the
effect of the vibration and gave one point on the flow
curve of the concrete under vibration. The concrete
was then remixed, replaced in the bowl, and
measurements repeated at a lower impeller speed to give
another point on the flow curve. In this way four
points were obtained and the flow curve constructed.
       The curve for the unvibrated concrete was the
usual straight line described by the equation

$$T = g + hN \qquad (1)$$

where T is the torque at speed N, g is the intercept on
the torque axis and h is the reciprocal slope; g and h
are measures of the yield value and plastic viscosity
respectively. The flow curve of the vibrated concrete
was not linear and appeared to start at, or near, the
origin as in Fig.1. Although experiments were carried

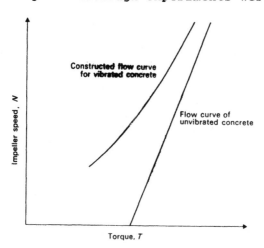

Fig.1. Effect of vibration on concrete flow curve.

out at several vibrator speeds (i.e. nominal frequencies) and amplitudes, no general conclusions could be drawn because the waveform of the crude vibrator used was so complicated, but it was clear that the method was well worth pursuing when a more controllable vibrator could be available.

In the meantime, it was possible to carry out experiments on a smaller scale by mounting a 100mm bowl on a small electromagnetic vibrator and using an impeller scaled down in the same ratio (2/5). The impeller was driven by the headgear of a Ferranti-Shirley viscometer which provided both the means of speed control and of torque measurement. Dimond (16) obtained results on Newtonian oils, pseudoplastic carboxymethyl cellulose solutions, dilatant cornflower pastes, laponite suspensions and cement pastes and showed

1)   There was no effect of vibration on the rheological properties of the first three materials i.e those materials that had no yield value.

2)   There was a marked effect on the yield values of the laponite suspensions and the cement pastes

3)   When vibration did have an effect, that effect was immediate and was also immediately and instantaneously reversible, that is, switching on the vibrator caused an immediate drop in torque and switching it off caused the torque to return immediately to its original value.

Experiments on Cement Pastes

The apparatus used by Dimond was later (17) used for experiments on aqueous pastes of ordinary Portland cement of water/cement ratios 0.28, 0.30, and 0.32. Full flow curves (i.e. both up and down curves) with a maximum speed of 200 rpm and a total sweep time of 136 secs were obtained at frequencies of 25, 50, 75 and 100 Hz, each at four amplitudes corresponding to accelerations of 1.5, 2.25, $3.0g_r$ and one higher acceleration whose actual value depended on the frequency being used. The paste was prepared in a Kenwood mixer operating at its lowest speed, then 10 mins after the first addition of water to the cement a flow curve (i) of the unvibrated paste was obtained, followed immediately by a curve (ii) obtained under the chosen vibration conditions, followed by a second flow curve (iii) without vibration.

It was found that curve (i) showed considerable hysteresis indicating that appreciable structural breakdown occurred. Curves (ii) and (iii) (during and after vibration) showed little or no hysteresis and the former was well to the lower torque side of the latter which, in turn, was close to or overlapped the down curve of curve (i). Thus the application of vibration reduced markedly the apparent viscosity of the paste, but the phenomenon was reversible.

The vibrated down curves were of the power law pseudoplastic type, that is they obeyed the equation

$$T = CN^b \qquad (2)$$

and the constants C and b were themselves related by the equation

$$C = pe^{-rb} \qquad (3)$$

so the vibrated curve could be described in terms of only one parameter b, for all test conditions

$$T = pe^{-rb} N^b \qquad (4)$$

with p and r as numerical constants equal to 91 and 5.6 respectively. b increased with decreasing frequency, increasing amplitude and increasing water/cement ratio. (Note that increase in b produces decrease in T).

The curves after vibration also fitted a power law of the same form as equation (2) and moreover the constants C_0 and b_0 fitted the same relationship as that between C and b of equation (2) so it could be written in the form

$$T_0 = pe^{-rb_0} N^{b_0} \qquad (5)$$

It was found that at each frequency $(b - b_0)$ was simply proportional to amplitude A, with the points for all W/C ratios fitting on the same line, and that the constant of proportionality was itself proportional to frequency F so

$$b - b_0 = \text{constant} \times FA = \text{constant} \times \text{velocity } v \qquad (6)$$

This relationship is confirmed by Figure 2 which shows the points from 48 separate experiments.

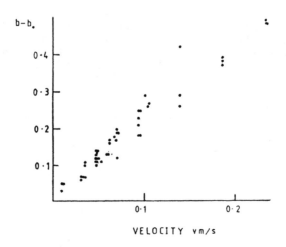

Fig.2. (b - b₀) as a function of v (cement pastes).

By substituting in equations (2) and (4) it follows that

$$T / T_o = K N^{k\,v} \qquad (7)$$

where K and k are constants for the particular experimental set-up but are independent of frequency, amplitude and water/cement ratio. Thus, the fractional drop in torque at a given speed depends only on the maximum velocity of the vibration and, as stated previously, it is reversible.

Flow Curves of Fresh Concrete

When an electromagnetic vibrator of sufficient capacity became available experiments to determine the flow curves of vibrated and unvibrated concrete were (18) carried out. The apparatus shown in Figure 3, was essentially the same as that in the preliminary work described earlier, but the crude mechanical vibrator was replaced by a Derritron electromagnetic vibrator whose waveform was close to a sine wave, and whose amplitude and frequency could be controlled independently. The peak thrust was 70 kN so a load of 70 kg could be given accelerations up to 10g.

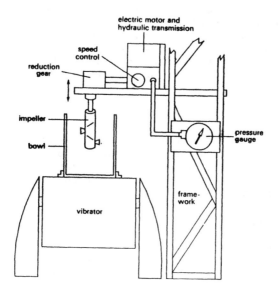

Fig.3. Experimental set-up.

For two mixes, each of 75mm slump, flow curves were obtained by the method already described, except that for the vibrated curves the number of points was increased from 4 to 7. Tests, each of which was completed in about 20 minutes, were carried out according to a strict timetable at six frequencies ranging from 15 to 100 Hz and at amplitudes corresponding to accelerations of 2½, 5, 7½, and 10g_r, with enough replications for a check on reproducibility.

The vibrated curves were similar qualitatively to those obtained in the preliminary experiments but, because of the better definition of vibration parameters, an attempt at a quantitative treatment was now justified. It was found that the vibrated curve could not be satisfactorily described as one that started from the origin and approached the non-vibrated curve as asymptote, but the results did fit fairly well to a power law curve

$$T_v = A_v . N^{B_v} \qquad (8)$$

where the suffix v indicates results obtained under vibration. It is not very convenient to compare this directly with the linear curve of the unvibrated material but it can be shown that the latter can also, over the restricted range of values of N considered, be

quite accurately represented by a similar power law
equation. The effect of applying vibration can then be
studied by considering the changes in the values of A
and B and in particular of the ratio A_v/A and the
difference $B_v - B$, because the ratio of the apparent
viscosities of the vibrated and unvibrated concrete is
given by

$$T_v \; / \; T = (A_v \; / \; A).N^{B_v - B} \qquad (9)$$

It was found that $A_v \; / \; A$ could be best represented as
an exponential function of acceleration g_r (in
gravitational units) and $B_v - B$ as an exponential
function of velocity v. Substituting in equation (11)
gave the complicated expression

$$T_v \; / \; T = K_1 \; \exp \; (\; - \; K_2 \, g).N^{K_3(1-\exp(-k_4 v))} \qquad (10)$$

where the values of K are known numerically from
experiment and differ for the two mixes. The values of
A and B for the unvibrated curve may be calculated from

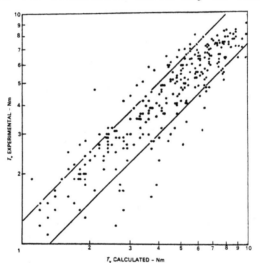

Fig.4. Comparison of experimental and calculated values
of T_v .

the values of g and h for the (correct) linear
relationship and may then be substituted in equation
(10) so that a comparison may be made between the
calculated and experimental values of T_v. The result
is shown in Figure 4 for the first mix and 80% of the

76 points lie within ±25% of the line of equality; for
he second mix, 82% of 130 points lie within such a
ange.

The final equations obtained by the power law
reatment are so complicated that it is not easy to see
he effect of altering vibration parameters and, in
ddition, although the agreement shown in Figure 4 is
ncouraging, the spread of results is rather large. It
s therefore worth taking advantage of the fact that at
ery low shear rates the vibrated curve can be
atisfactorily approximated by a straight line through
he origin, that is, that over this restricted range
he vibrated concrete can be regarded as a Newtonian
luid which is characterised by a single constant. For
ach flow curve the best straight line from the origin
as drawn by eye and, for each, the slope S, which is a
easure of the fluidity, was obtained. It was found
hat S could be represented as a simple exponential
unction of amplitude A

$$S = S_0 \ (1 - e^{-cA}) \qquad\qquad (11)$$

nd that for each mix the constant c was a simple
linear function of frequency F of the form

$$c = k \ (F - F_0) \qquad\qquad (12)$$

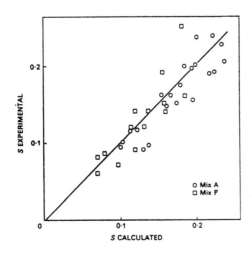

Fig.5. Comparison of experimental and calculated values
of S.

where F_0 is a small threshold frequency. Substitution gives

$$S = S_0 (1 - \exp (- Ak (F - F_0))) \qquad (13)$$

The values of S_0, k and F_0 are known from the experimental results.

The quality of agreement between the values of slope (fluidity) obtained from equation (13) and those from the experimental flow curves is shown in Figure 5. Provided F is appreciably greater than F_0 the exponent reduces to AkF, that is, a constant multiplied by velocity, so that, once again, the effectiveness of vibration is a simple function of velocity only.

Vertical Pipe Experiments

The finding that at low shear rates concrete under vibration behaves as a Newtonian fluid presents the possibility of a simpler method of investigation, which was exploited (19) in the apparatus shown in the diagram in Figure 6. A vertical pvc pipe, 700mm long with internal diameter 100mm, was mounted centrally inside a 356mm dia. cylindrical bowl that was firmly clamped to the vibrating table. The lower end of the

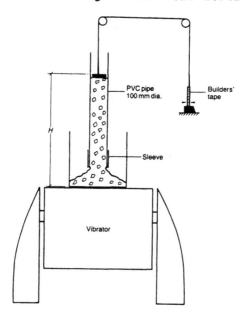

Fig.6. Experimental set-up.

pipe was 100mm above the base of the bowl but could be
lengthened to touch it by means of a sleeve. In use,
the pipe was filled with the concrete under test while
the sleeve was in its lowered position, and then the
sleeve was raised so that the concrete was unsupported
but remained in position because of its low
workability. The vibrator was then switched on at the
chosen amplitude and frequency and the downward
movement of the concrete was monitored by reading the
measuring tape against a fixed index, the tape being
pulled by a string connected to a light piston placed
on top of the concrete.

Six different mixes were investigated at all
combinations of frequencies 200, 160, 120, 100, 80, 75,
60, 50, 35, and 15 Hz and amplitudes corresponding to
accelerations of 0.85, 1.27, 2.54, 3.31, 5.08, 6.36,
7.63 and 8.90 g_r except when dH/dt was too fast to
measure.
If the concrete under vibration behaves as a
Newtonian fluid under the rates of shear applying, the
rate of flow can be expected to be simply proportional
to the hydrostatic head which, in turn, will be simply
proportional to the height H if either all the concrete
in the pipe is sufficiently affected to have zero yield
value, or if the drag at the concrete-pipe interface is
negligible or proportional to the area of contact.
That is

$$dH/dt = - bH \qquad (14)$$

It was found that in all cases the relationship between
ln H and t was linear over at least the first 15 secs,
thus confirming the validity of equation (14). At
constant frequency F, b was linearly related to
amplitude A

$$- b = m(A - A_0) \qquad (15)$$

where m and A_0 are constants and the latter represents
some small threshold amplitude below which vibration
has no effect. Both these constants were found to be
fairly simple functions of frequency

$$- K_5 m = \ln(1 - F/F^*) \qquad (16)$$

and $\qquad A_0 = K_6 + K_7 (F/F^*)^{K_8} \qquad (17)$

where F* is an upper limiting frequency for which m
becomes infinite, and K_1 are constants whose values are
found experimentally. Combining these equations gives

$$b = K_9 \ln (1 - F/F^*) \times (A - K_6 - K_7 (F/F^*)^{K_8}) \quad (18)$$

The values of b obtained from this equation for one of
the mixes are shown plotted against experimental values
in Figure 7. The agreement is good.

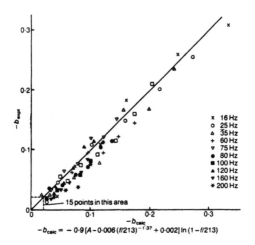

$$-b_{calc} = -0.9[A-0.006(f/213)^{-1.37} + 0.002] \ln (1 -f/213)$$

Fig.7. Comparison of experimental and calculated values
of b for one mix.

Comparison of results

If it is justifiable to treat vibrated concrete as a
Newtonian fluid at low shear rates, and if the
equations obtained from the results of the two sets of
experiments are valid, then the quantities S and b
should, for the same mix, be simply related. For
practical reasons the same mixes were not used in the
two types of experiment but there is one case where
mixes differing in water content only were used. For
these two it is still reasonable to expect a simple
relationship although with some departure from
linearity. A plot of S against b gave a good
correlation but the points for a frequency of 15 Hz
were off the line. This was clearly because at this
frequency the value of the threshold frequency F_0 (16
Hz) had a large effect and it was recognised that there
was some doubt about its actual value. The line for S
as a function of F was therefore recalculated with the

imposed condition that it must pass through the origin
(i.e. F_0 = 0) and the resulting equation became

$$S = {}^1/_4 \ (\ 1 \ - \ \exp(\ -0.03 \ AF)) \qquad (19)$$

A plot of this function against the values of b showed
very good correlation, as illustrated in Figure 8,
although, as expected there was some departure from
linearity.

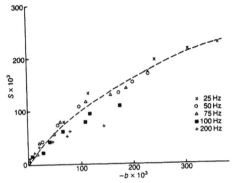

Fig.8. Relationship between measures of fluidity from
flow curves, S, and from pipe experiments, b.

Discussion

The excellent correlation shown in Figure 8 is strong
support for the arguments given and for the contention
that fresh concrete under vibration behaves as a simple
Newtonian at low shear rates, if certain threshold
vibration conditions are exceeded. Provided amplitude
and frequency applied differ appreciably from the
threshold and limiting values, the expressions for S
and b both reduce to simple functions of velocity or,
in other words, the maximum velocity of the vibration
is the important parameter in determining concrete
behaviour, and this finding is in agreement with the
results of the cement paste experiments.
    There was some evidence that the limiting
frequency F^* had a value nearer to 500 Hz than to the
values (around 250 Hz) used in the equations so
equation (18) was recalculated to allow for this. This
new equation for b is shown plotted as a function of
frequency at several constant values of acceleration
and also at several constant values of velocity, in
Figure 9. It can be seen that b is almost constant at

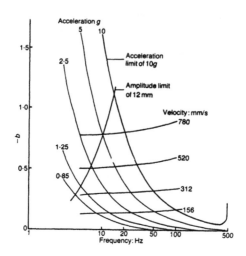

Fig.9. Fluidity expressed as b calculated from modified
equation (18) as a function of frequency at constant
values of acceleration and velocity.

constant velocity, but drops off rapidly with frequency
at constant acceleration until the limiting frequency F
= F* = 500 Hz is approached closely when theoretically
it becomes very large. Practical limits are imposed by
the characteristics of the equipment. In this
particular case, acceleration cannot exceed 10g and,
even at low frequencies, amplitude cannot exceed 12mm.
The result is to produce a cusp-shaped working region
and the best conditions are at the apex of the cusp
i.e. at 14 Hz. The shape of the cusp, and therefore
the position of the apex, depend on the characteristics
of the equipment but it may be pointed out that the
value of 14 Hz indicated here is much lower than the
nominal frequencies associated with the commercial
vibrating tables in use in industry.

References

1) Institution of Civil Engineers & Institution of
   Structural Engineers. The Vibration of Concrete
   Report of a Joint Committee. 1956. pp.64.
2) Murphy, W.E. A survey of post-war British research
   on the vibration of concrete. London. Cement
   and Concrete Association. 1964. pp.25. Technical
   Report TRA 382.

) ACI Committee 309. Behaviour of fresh concrete during vibration. Journal of the American Concrete Institute Proceedings. 78(1), 1981, 36-53.

) American Concrete Institute. Consolidation of Concrete Ed. S.H. Gebler. Detroit. 1987. Special Pub'n SP-96, pp250.

) Whiffin, A.C. & Smith, R.T. The measurement of vibrations in freshly placed concrete. Magazine of Concrete Research, July 1950, 39-46.

) Kirkham, R.H.H. The compaction of concrete slabs by surface vibration: first series of experiments. Magazine of Concrete Research Dec.1951, 79-91.

) Venkatramaiah, S. Measurement of work done in compacting a known weight of concrete by vibration. Magazine of Concrete Research Jan. 1951, 89-96.

) Cusens, A.R. The influence of amplitude and frequency in the compaction of concrete by vibration. Magazine of Concrete Research. Aug. 1958 79-86.

) L'Hermite R. The rheology of fresh concrete and vibration Paris. CERILH 1948. Publication Technique No.2.

10) Tattersall, G.H. The rheology of cement pastes, fresh mortars and concretes. M.Sc Thesis. University of London. 1954.

11) Tattersall, G.H. Structural breakdown of cement paste at constant rate of shear. Nature 175 (4447), 166, 1955.

12) Tattersall, G.H. The rheology of Portland cement paste. British Journal of Applied Physics 6 (5), 165-167. 1955.

13) Lassalle, a. & Legrand, C. Changes in the rheological behaviour of cement paste with distance from a vibrating source. Materiaux et constructions 13, 1980, 115-124.

14) Tattersall, G.H. & Bloomer S.J. Further development of the two-point test for workability and extension of its range. Magazine of Concrete Research 31 (109).

15) Tattersall, G.H. & Banfill, P.F.G. Rheology of Fresh Concrete London, Pitman. 1983. pp.356.

16) Dimond, C.R. Unpublished internal report. Department of Building Science, University of Sheffield. 1980. Reported in ref.15. pp.301-302.

17) Tattersall, G.H. The effect of vibration on the flow properties of cement paste. Proceedings of Conference on Mechanics and Technology of Composite Materials Varna. Bulgaria, 1985. Sofia, Bulgarian Academy of Sciences. 419-422.

18) Tattersall, G.H. & Baker, P.H. The effect of
    vibration on the rheological properties of fresh
    concrete. Magazine of Concrete Research <u>40</u> (143)
    June 1988 79-89.
19) Tattersall, G.H. & Baker, P.H. An investigation on
    the effect of vibration on the workability of
    fresh concrete using a vertical pipe apparatus.
    Magazine of Concrete Research <u>41</u> (146), Mar.
    1989, 3-9.

34 RHEOLOGY OF FRESH CONCRETE UNDER VIBRATION

S. KAKUTA
Akashi Technological College, Akashi, Japan
T. KOJIMA
Ritsumeikan University, Kyoto, Japan

Abstract
This paper describes the rheological estimation of the fresh concrete under vibration. In this study rheological behavior of fresh concrete was tested by a rotating fan type rheometer with a table vibrator at the bottom of the container. This rotating fan type rheometer was developed from Tattersall's Two-point workability tester.
It was indicated that the flow behavior of fresh concrete had a linear relationship between the torque of impeller T and the revolution speed N in the static test. When vibrated, the linear relation was, however, changed into an exponential relation in which yield value disappeared. This flow was considered to be a quasiviscous flow. The flow curve could be expressed as an exponential equation $T = p \, N^q$, coefficients of which could be obtained by the test. The measured values of the coefficients were depend on the mix proportion and the chemical admixture.
It was observed that there was a region in some flow curves where measured torque became higher than that without vibration. It may be considered that this phenomenon was caused by sedimentation of coarse aggregate due to vibration.
Keywords: Rheology, Vibration, Rotating Fan Type Rheometer, Quasiviscous Flow, Sedimentation.

1 Introduction

Mechanical vibrators are commonly used as a method to densify fresh concrete. When the vibration is applied to freshly placed concrete even, low slump concrete flows almost as viscous liquid. In such a case of a heavily reinforced concrete and a complicated-shaped structure,the use of vibrator becomes important not only for the consolidation but also for the fluidization of concrete.
There are only a few investigations on the flow behavior of fresh concrete under vibration[1][2]. It is simply because it is not easy to establish the effective experimental method to obtain the rheological quantities under vibration. In this study a rotating fan type rheometer was modified to be able to use under vibration by using a table vibrator at the bottom of the rheometer. The experimental results showed that the fresh concrete under vibration behaved as a quasiviscous flow.

2 Experiment

2.1 Materials
Five concrete mixes were examined Ordinary Portland Cement, river sand for fine aggregate and crushed hard sandstone for coarse aggregate were used in each mixes.
A hydroxyethylcellulose(HEC) based admixture and AE water reducing agent(AE WRA) were used for a segregation controlling admixture. Water cement ratio and sand aggregate ratio were kept in constant values of 50% and 48% respectively. Used materials and there contents are summarized as follows:

Water Content   190,200,210 and 218 kg/m^3
Cement content   380,400,420 and 436 kg/m^3
Superplasticizer   0.006 per cement content
Segregation controlling admixture
                   0.0075 per cement content(HEC)
                   0.025 per cement content(AE WRA).

2.2 Test apparatus
A rotating fan type rheometer similar to that proposed by Tattersall[1] was used. A table vibrator, such as that used in Japan, was used at the bottom of container. The vibration frequency was 29.1 Hz. The test set-up is shown in Fig 1.

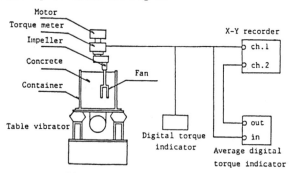

Fig.1.Experimental set-up.

The measured vertical accelerations of container at the middle level from the bottom were 2.00g,2.50g,2.67g and 3.17g for vibration number 1, 2, 3 and 4 respectively. Two pore water pressure gauges were set up in the middle of container at two levels of 7cm and 17 cm from concrete surface. Test conditions and test procedures are as follows:

Concrete temperature   20 ± 1˚ C.
Mixing time   30 sec(without water) and 180 sec
                                (with water and admixtures)
Average torque reading time   10 sec.
Impeller speed   0.005,0.20,0.35,0.50,0.80,1.10 and
                                1.14 revolutions per second.

3 Results and discussions
Fig.2 shows the flow curves of plain concrete with water content of 190kg/m^3 both under and without vibration. The structural break down

is occurred by vibration so that the up-curve of flow curve almost agreed with the down curve and the yield value disappeared. Fig.3 shows typical down curves.

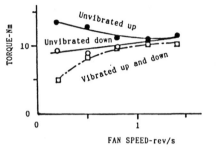

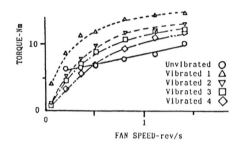

Fig.2.Flow curve of concrete under and without vibration.

Fig.3.Typical flow curves of concrete under vibration.

A quasiviscous equation form is applicable to a flow curve of concrete under vibration:

$$T = p\ N^q \qquad (1)$$

where p is the coefficient on apparent viscosity and q is the power number on apparent viscosity index. Fig.4 shows relationships between coefficient p and power number q for various water contents of concrete.

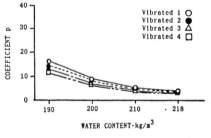

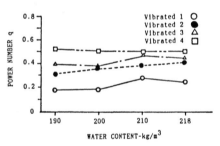

Fig.4.Relationship between p and q.

Stress of fresh concrete in a container is expressed by the following equation

$$s = s' + u \qquad (2)$$

where s is stress of fresh concrete in a container, s' is effective pressure and u is pore water pressure. Pore water pressure changes when repeated shear stress is applied to fresh concrete. Fresh concrete in the container may be assumed to be in undrained condition. If a pore water pressure is increased by cyclic stress, then an effective stress decreases. The liquefaction occurs simultaneously when the effective stress becomes to zero. Typical relationships between excess pore water pressure and depth from concrete surface for several vibration times measured in the case of unit water content of $190 kg/m^3$ under vibration 1 are shown in Fig.5. It indicates the differences of water pressure between surface and inner concrete such as that the hydraulic gradient was occurred by vibration. From this figure the pore water rising speed to the concrete surface can be estimated. A flow curve of a non-segregation-in-water concrete produced by a segregation controlling admixture is shown in Fig.6. In high speed range of Fig.3, measured torque under vibration became greater than that without vibration. It is because the sedimentation of solid particles near the bottom of container lead to increase apparent torque. On the other hand, there is no evidence in Fig.6. It indicates that the segregation did not occur in non-segregation-in-water concrete.

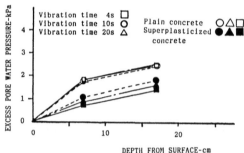

Fig.5.Excess pore water pressure of inner concrete under vibration.

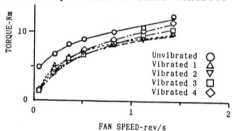

Fig.6.Flow curve of non-segregation-in-water concrete under vibration.

4 Conclusions

(1) The flow curves of under vibration show quasiviscous or pseudoplastic fluid behavior. The non-Newtonian factors are appropriate in evaluation of flow of concrete under vibration.

(2) In some of the flow curves concrete under vibration, measured torque became greater than that without vibration in high speed range, but it could be observed in non-segregation-in-water concrete.

(3) Flow behavior under vibration is related to the acceleration and the pore water pressure. Excess pore water was observed when the concrete vibrated.

5 References

[1]Tattersall,G.H. and Banfill,P.F.G.(1983)The Rheology of Fresh Concrete,Pitman Advanced Publishing Program,London.
[2]Tattersall,G.H. and Baker,P.H.(1988)The effect of vibration on the rheological properties of fresh concrete,Magazine of Concrete Research,**40**,143,79-89.

35 DIPHASIC SEMI-EMPIRIC MODEL FOR THE RHEOLOGICAL BEHAVIOUR OF VIBRATED FRESH CONCRETE

M. BARRIOULET and C. LEGRAND
Laboratoire Matériaux et Durabilité des Constructions,
INSA-UPS, Toulouse, France

Abstract
The propounded model aims at describing the steady flow of
vibrated fresh concrete. One phase is the cement paste, the
rheological behaviour of which is simplified by a newtonian law ; the other phase is the aggregate, the action of
which is represented in the concrete flow by the influence
of the forced translation of a solid and long single cylinder (radius R_1) on the flow of the cement paste under a
constant pressure gradient in the annular space between the
last cylinder and a coaxial pipe (radius R_2).
$\alpha = R_1{}^2/R_2{}^2$ traduces the volumic concentration Γ of aggregate in the concrete and the model gives the average velocity v of the flow.
Theoretical curves representing v versus Γ are compared
with experimental results proceeding from the steady flow
of various concretes through a nozzle with vertical axis.
The studied parameters are : the aggregate concentration Γ,
the shape and the density of grains.
Keywords : Two-phases model, Vibrated Fresh Concrete, Flow,
Heterogeneity.

1 Introduction

The heterogeneity of fresh concrete is an important hin-
drance to the global mechanical characterisation of this
material. In particular, it makes every classical rheological measurement, especially the viscometric one, inadequate
(Barrioulet and al, 1978).
A simplification of the problem consists in reducing
the heterogeneity to a two-phases material composed of a
granular phase and an interstitial paste.
Studies of the flow of vibrated fresh concrete (Bombled, 1964) (Barrioulet and al., 1986) show that the macroscopic behaviour of such material does not result only from
the superposition of that of the paste and of that of the
aggregates. Moreover, mechanical interactions between
phases take place, especially mass effects of grains on the
paste and lubrication by the latter of the intergranular

actions.

The purpose of this paper is to describe an analogic two-phases semi-empiric model of the flow of vibrated fresh concrete. We have tried to build this model as simple as possible, our aim in this way being only a first approach in the understanding of the phenomena.

2 Analogical model

2.1 Hypothesis

The movement of the whole granular phase is represented by that of a solid cylinder, the unit weight of which is that of the aggregate (γ_g). This cylinder, the length of which is infinite and the radius R_1, is compelled to follow the direction 1 in translation and without solid friction (Fig. 1)

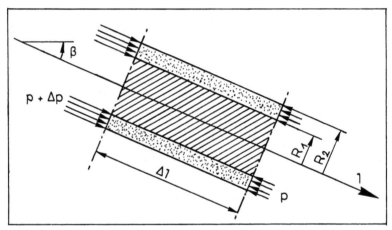

Fig. 1. Analogical model

The cement paste flows in the annular space between the moving cylinder and a fixed pipe of radius R_2, under the actions of its own weight and of a pressure gradient
$- dp/dl$.

As we will specify later, the parameter $\alpha = R_1^2/R_2^2$ is a function of the volumic concentration Γ of the aggregate in the concrete and R_2 depends on the geometrical characteristics of the flow.

In order to simplify the calculations, we suppose that:

- the rheological behaviour of the vibrated paste is newtonian, while we know it is pseudoplastic (Legrand, 1972)
- the flow occurs under a constant pressure gradient
($- dp/dl = - \Delta p/\Delta l = $ Cte)
- there is no slippage on the sides.

2.2 Calculation of the rate of flow

Let $P = p + \gamma_p\, l \sin \beta$

where p = pressure

       γ_p = unit weight of the paste

Differentiating,

$$dP/dl = (dp/dl) + \gamma_p \sin \beta$$
$$= (\Delta p/\Delta l) + \gamma_p \sin \beta = \Delta P/\Delta l = Cte$$

The fluid mechanics theory gives :

$$\frac{dP}{dl} = \frac{\eta}{r} \frac{d}{dr} \left(r \frac{du}{dr} \right) \qquad (1)$$

where η = newtonian viscosity of the vibrated paste

       u = velocity in the direction l

       r = radial coordinate

The general solution of eq. (1) is :

$$u = Ar^2 + B\ln r + C \qquad (2)$$

According to limit conditions, we get :

$$A = (1/4\eta)(\Delta P/\Delta l) \qquad B = \frac{U_0 - (1/4\eta)(\Delta P/\Delta l)(R_1{}^2 - R_2{}^2)}{\ln R_1/R_2}$$

$$C = - AR_2{}^2 - B \ln R_2$$

where U_0 = velocity of the solid cylinder

Hence, let us write the solid cylinder equilibrium :

$$\pi R_1{}^2 \; \Delta l \; \gamma' \sin \beta - 2 \pi R_1 \; \Delta l \; \tau(R_1) = 0$$

where

     γ' = submerged unit weight of the cylinder

     $\tau = - \eta\, du/dr$ (viscous friction)

So, $R_1 \; \gamma' \sin \beta + 2\eta \; (du/dr)_{r=R_1} = 0 \qquad (3)$

Now, according to eq. (2)

$$du/dr = 2Ar + B/r$$

Substituting in eq. (3)

$$R_1 \; \gamma' \sin \beta + 4\eta \; R_1 A + 2\eta \; B/R_1 = 0$$

According to the expressions of A an B, we can deduce

$$U_0 = \frac{1}{4\eta}\frac{\Delta P}{\Delta l}(R_1{}^2 - R_2{}^2)(1 - \frac{2R_1{}^2}{R_1{}^2 - R_2{}^2}\ \ln R_1/R_2)$$
$$- \frac{R_1{}^2}{2\eta}\gamma'\ \sin\beta\ \ln R_1/R_2 \tag{4}$$

The rate of flow of the paste is

$$Q_p = 2\pi \int_{R_1}^{R_2} ur\ dr$$

After calculation, we get

$$Q_p = \frac{\pi}{8\eta}\frac{\Delta P}{\Delta l}\frac{R_2{}^2 - R_1{}^2}{\ln R_1/R_2}\left[R_1{}^2(1 - \ln R_1/R_2) - R_2{}^2(1 + \ln R_1/R_2)\right]$$
$$+ \pi U_0\frac{1}{2\ln R_1/R_2}\left[R_1{}^2(1 - 2\ln R_1/R_2) - R_2{}^2\right]$$

The solid rate of flow of the inner cylinder is

$$Q_g = \pi R_1{}^2 U_0$$

The total rate is $Q = Q_p + Q_g$; so,

$$Q = \frac{\pi}{8\eta}\frac{\Delta P}{\Delta l}\frac{R_2{}^2 - R_1{}^2}{\ln R_1/R_2}\left[R_1{}^2(1 - \ln R_1/R_2) - R_2{}^2(1 + \ln R_1/R_2)\right]$$
$$+ \pi U_0\frac{1}{2\ln R_1/R_2}(R_1{}^2 - R_2{}^2) \tag{5}$$

2.3 Average velocity of the flow
Let the average velocity v of the flow be

$$v = Q/\pi R_2{}^2$$

According to eq. (5), we get :

$$v = \frac{1}{8\eta}\frac{\Delta P}{\Delta l}\frac{R_2{}^2 - R_1{}^2}{R_2{}^2}\cdot\frac{1}{\ln R_1/R_2}\left[R_1{}^2(1 - \ln R_1/R_2) - R_2{}^2(1 + \ln R_1/R_2)\right]$$
$$+ \frac{U_0}{2\ln R_1/R_2}\frac{R_1{}^2 - R_2{}^2}{R_2{}^2} \tag{6}$$

If we make $R_1 = 0$, v becomes the average velocity v_p of the paste flowing alone in a pipe of radius R_2. Eq. (6) gives

$$v_p = -(1/8\eta)(\Delta P/\Delta l)R_2{}^2 \tag{7}$$

Introducing v_p and $\alpha = R_1{}^2/R_2{}^2$ in eq. (6)

$$v = v_p\frac{\alpha - 1}{\ln\alpha}\left[\alpha(2 - \ln\alpha) - (2 + \ln\alpha)\right] + U_0\frac{\alpha - 1}{\ln\alpha} \tag{8}$$

U_0 is given by eq. (4) ; if
$v_g = (\gamma'\ \sin\beta/4\eta)\ R_2{}^2$ is a conventional velocity, eq. (4) can be written

$$U_0 = -2 v_p (\alpha-1 - \alpha \ln \alpha) - v_g \alpha \ln \alpha$$

Substituting in eq. (8), we finally get

$$v = v_p(\alpha-1)^2 - v_g\alpha(\alpha-1) \qquad (9)$$

2.4 Adaptation of the model to a real flow

As we have said at the beginning, α is a function of the volumic concentration Γ of the aggregate in the concrete ; we have tried a linear correlation like
$$\alpha = k\Gamma$$
Let $\bar{v}_p$ be the real average velocity of the single paste ($\Gamma=0$) and $\bar{v}_g$ the conventional velocity in the real flow ; we write
$$\bar{v}_p = fv_p$$
$$\bar{v}_g = fv_g$$
where $f = R^2/R_2^2$, R beeing a parameter of adaptation, function of the geometry of the flow.

Substituting in eq. (9), we obtain the average velocity $\bar{v}$ of the real flow :

$$\bar{v} = k^2(\bar{v}_p - f v_g)\Gamma^2 - k(2\bar{v}_p - f v_g)\Gamma + \bar{v}_p \qquad (10)$$

3 Experimental validation

3.1 Experimental device

According to the constant pressure gradient hypothesis, we have realized a flow under constant head by the help of an apparatus, the description of which has been already publicated (Barrioulet and al., 1982). Briefly, it consists (Fig. 2) in a vertical pipe ($\beta = \pi/2$) fed permanently to the top by the help of a concrete reservoir and of an over-flow system. The whole device is fixed on a vibrating table. The fresh concrete flows through a tronconic nozzle and falls on a recording scale that allows the measurement of the rate of flow. Then, it is easy to determine the average velocity of the flow through the nozzle.

3.2 Composition of concretes

The only parameters we have varied have been the concentration, the nature and the shape of the aggregates (all of them presented approaching granulometries).

The cement paste was simulated by a calcite filler mix (to avoid rheological evolution during time) the volumic solid concentration of which was 0.570. Viscometric measurements realized on the vibrated paste allow us to choose, with a good approximation, $\eta = 10$ Pa.s. The real average velocity $\bar{v}_p$, directly measured with the apparatus is 4.1 cm.s^{-1}.

We have used three kinds of crushed aggregates
- limestone $\qquad \qquad \gamma_g = 27.2$ kN.m^{-3}
- barited rock $\qquad \qquad \gamma_g = 36.4$ kN.m^{-3}

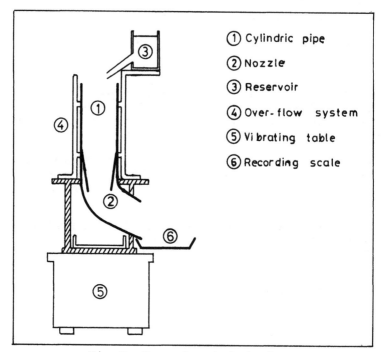

Fig.2. Experimental device

1. Cylindric pipe
2. Nozzle
3. Reservoir
4. Over-flow system
5. Vibrating table
6. Recording scale

- hardened cement paste γ_g = 21.4 kN.m^{-3},
and three kinds of rounded aggregates
- glass γ_g = 25.6 kN.m^{-3}
- expanded clay γ_g = 15.0 kN.m^{-3} (saturated)
- polystyrene γ_g = 0.2 kN.m^{-3}

3.3 Results with crushed aggregates
a - Limestone

We find $\bar{v}_g = fv_g = (\gamma'/4\eta)R^2$ = 4.3 R^2 cm. We have chosen
a semi-empirical way in adjusting with experimental re-
sults, for instance those obtained with limestone (fig. 3,
curve 1) ;

we get   R = 1.56 cm
       and   k = 2.0

b - Barited rock

Using the same values for R and k than before, the calculation gives $\bar{v}_g$ = 16.1 cm.s^{-1}
and $\bar{v}$ = - 48.0 Γ^2 + 15.8 Γ + 4.1 (fig. 3, curve 2).

c - Hardened cement paste

By the same way, we obtain $\bar{v}_g$ = 7.0 cm.s^{-1}
and $\bar{v}$ = - 11.6 Γ^2 - 2.45 + 4.1 (fig. 3, curve 3).

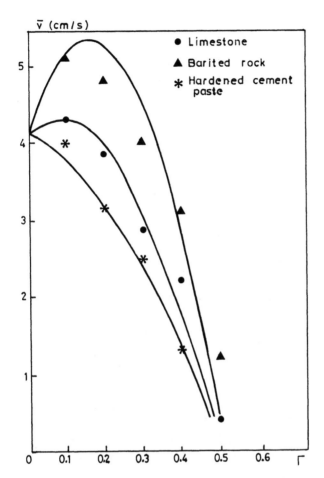

Fig. 3. Theoretical curves and experimental points
obtained with crushed aggregates

3.4 Results with rounded aggregates

We have found that the preceeding values for R and k are
not convenient when the shape of the grains changes. So, we
have chosen to adjust these parameters for rounded aggregates with the experimental results obtained with glass
balls.

a - Glass

We find $\bar{V}_g = 3.9\ R^2$ cm

R = 1.70 cm

k = 1.58 (fig. 4, curve 1)

b - Expanded clay

The calculation gives $\bar{V}_g = 3.6\ cm.s^{-1}$

and $\bar{V} = 1.25\ \Gamma^2 - 7.27\ \Gamma + 4.1$ (fig. 4, curve 2)

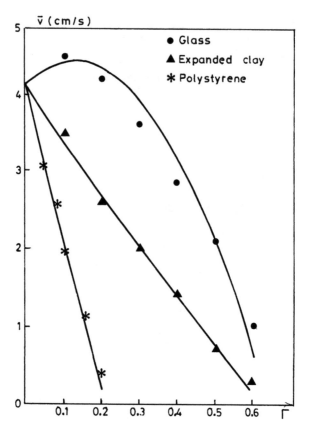

Fig. 4. Theoretical curves and experimental points
obtained with rounded aggregates

c - Polystyrene
$\bar{v}_g = - 7.06$ cm.s^{-1}
$\bar{v} = 27.9\ \Gamma^2 - 24.1\ \Gamma + 4.1$ (fig. 4, curve 3)

4 Conclusions

In spite of its roughness, this model gives surprising good
results, chiefly if we take into account the extreme variety of aggregates we have used.

Indeed, it is now indispensable to try validations with
other concretes and other kinds of flows. Nevertheless, it
already allows us to understand better what the mechanical
interactions between paste and aggregate during the flow of
fresh concrete are.

References

Barrioulet, M. and Legrand, C. (1978) L'hétérogénéité, caractère essentiel de l'écoulement du béton frais vibré. Silicates Industriels, 2, 29-34.

Barrioulet, M. and Legrand, C. (1982) Experimental study of fresh concrete vertical flow under constant load. Proceedings 3d Int. Conf. of Mech. and Techn. of Composite Materials, Varna, 400-403.

Barrioulet, M. and Legrand, C. (1986) Mise en évidence des interactions entre pâte interstitielle et granulats dans l'écoulement du béton vibré. Matériaux et Constructions, 19, 112, 273-278.

Bombled, J.P. (1964) Rhéologie du béton frais, Pub. tech. CERILH, n° 161.

Legrand, C. (1972) Les comportements rhéologiques du mortier frais. Cem. and Conc. Research, 2, 17-31.

PART EIGHT
DISCUSSION

GENERAL DISCUSSION
CONTRIBUTION ON CEMENT PASTE
STRUCTURE AND YIELD STRESS

S. KELHAM
Blue Circle Industries PLC, Greenhithe, UK

Many of the papers discuss cement paste rheology in terms of a flocculated system showing structural breakdown on shearing. Others refer to frictional forces between particles. I suggest here that a model based on frictional interactions can account for the observed behaviour of cement pastes, within the water/cement ratio range appropriate to concretes. Friction in dry powder systems leads to a yield stress but no breakdown under shear. The frictional properties of cement particles in a paste depend on the hydration products on the surfaces. Under shear these may be broken off, reducing friction, releasing trapped water and giving the observed breakdown behaviour. Vibration, as with powders, reduces the static friction and thus reduces the yield stress but does not damage the particle structure, giving a reversible effect. Extended shearing gives an increasing concentration of hydration products in the inter-particle fluid and an increased rate of reaction due to the exposure of unhydrated material. This will be observed as structural buildup in long timescale flow curve measurements.

Friction between hydrating particles can thus account for the observed behaviour. The relative importance of "floc" structure and particle-particle friction should be observable by carrying out shear yield stress experiments under controlled normal stress conditions. An increase in normal stress should increase the shear yield stress of a frictional system but reduce it in a flocculated system. It would be very interesting to test this hypothesis in a rheometer capable of taking measurements of normal stress under controlled shear stress conditions.

DISCUSSION CONTRIBUTION ON
CHAPPUIS (page 3) AND ON
GREGORY AND O'KEEFE (page 69)

M.F. BRAIN
Colas Building Products, Chester, UK

As a manufacturer of self-levelling smoothing compounds, we need to have low viscosity and low yield stress. Apart from simple viscometry and flow table tests, are there rheological test methods to characterise self-levelling?

AUTHOR'S REPLY (J. CHAPPUIS)

I do not know of any rheological test to characterise self-levelling apart from the flow table. However, as long as you know what you need in terms of viscosity and yield stress, I would recommend you to measure these parameters in a coaxial cylinder viscometer under conditions of low shear rates, in order to be relevant to the application conditions.

AUTHOR'S REPLY (T. GREGORY)

Using modern rheometers it is possible to design experiments in order to simulate any process. For example with self-levelling materials it is necessary to simulate the mixing by subjecting the material to a high shear rate and then to simulate the flow behaviour and structure buildup on placing by subjecting it to an oscillatory stress or by measuring the stress growth.

DISCUSSION CONTRIBUTION ON
OKAMOTO AND ENDOH (page 113)

A. TERENTYEV
Technical University of Riga, Latvia, USSR

In the theory of elasticity there are well known simple
relationships between Poisson's ratio and five elastic
moduli relevant to every kind of uniform loading for linear
isotropic elastic media (Terentyev and Kunnos, 1990).
Attempts have been made to obtain similar relationships for
materials displaying viscosity under some types of uniform
loading (Reiner, 1969). However, unsatisfactory results are
usually obtained because of the lack of serious scientific
grounds for obtaining the relationships, resulting from
application of simple analogies instead of strict proof.
The comparison between the columns in Fig. 1 show this.
Columns 1 and 2 give the rheological models assumed for a
material under triaxial and shear loading (Reiner, 1969).
Column 3 shows the rheological models and the corresponding
relationships between viscous moduli proposed by different
authors. Column 4 shows the true models and the
corresponding relationships between the rheological
parameters derived by our theory (Terentyev and Kunnos,
1990). Our theory can satisfactorily explain all the
difficulties and ambiguities associated with the work of
different authors on various kinds of uniform loading.

In the paper by Okamoto and Endoh three unfounded
hypotheses are used:

(i) The four element model under uniaxial stress is
extended to that under triaxial stress.

(ii) The elastic moduli under triaxial stress are equal
to those under uniaxial stress (equations 6 and 7).

(iii) The viscous moduli under triaxial stress are
expressed using Poisson's ratio and moduli relevant to
uniaxial loading, by analogy with the relationships known
for elastic materials (equations 9 and 10).

However all these defects can be easily removed with the
help of our theory. Moreover applying suitable models and
relationships from our theory enables us to eliminate the
significant difference between the calculated and measured
results in Figs. 13 and 14 of their paper, which are
primarily caused by the authors' incorrect hypothesis
(equation 6).

Of course, the authors would be unaware of our theory
during their work, as it has only just been published.

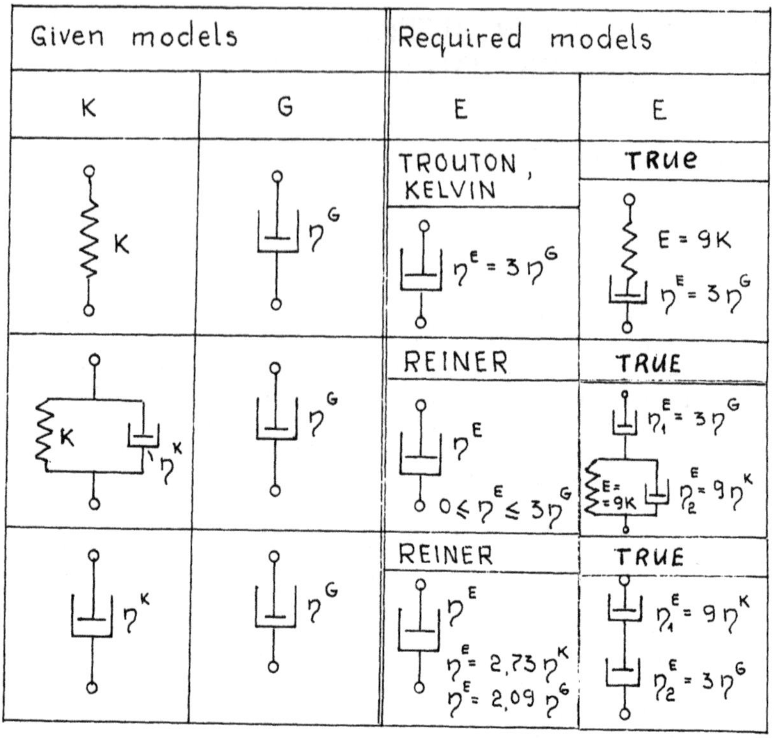

Given models		Required models	
K	G	E	E
K (spring)	η^G (dashpot)	TROUTON, KELVIN $\eta^E = 3\eta^G$	TRUE $E = 9K$, $\eta^E = 3\eta^G$
K, η^K	η^G	REINER η^E, $0 \leq \eta^E \leq 3\eta^G$	TRUE $\eta_1^E = 3\eta^G$, $E = 9K$, $\eta_2^E = 9\eta^K$
η^K	η^G	REINER η^E, $\eta^E = 2.73\eta^K$, $\eta^E = 2.09\eta^G$	TRUE $\eta_1^E = 9\eta^K$, $\eta_2^E = 3\eta^G$

Fig. 1. Models and relationships for viscoelastic materials (K - triaxial loading, G - shear, E - uniaxial loading).

However, I hope this discussion contribution may be useful in their future work.

References

Reiner, M. (1969) **Deformation, strain and flow.** Lewis, London.
Terentyev, A., Kunnos, G. (1990) Method of determining rheological parameters for linear viscoelastic medium under four kinds of uniform loading. **This volume, 91.**

Thank you for your kind suggestion to us.
We are afraid you may misunderstand Eq.(6), (7), (8), (9), (10) and (11)
in our report, as deriving viscoelastic 4-element model under three
axial stress from that under uni-axial stress. We are sorry we could
not sufficiently explain our intention because of the limited space of
our report. We would like to explain on this discussion in detail as
follows.
First, consider Maxwell element in generalized 4-element model as shown
in Fig.11. Decomposing the rate of strain $\{\dot{\varepsilon}_g\}$ into the deviatoric
component $\{\dot{\varepsilon}_g{}'\}$ and the volumetric component $\{\dot{\varepsilon}_{gm}\}$, we have

$$\{\dot{\varepsilon}_g{}'\} = \frac{C_{Gg}}{2}\{\dot{\sigma}'\} + \frac{1}{2\eta_{Gg}}\{\sigma'\} \tag{D-1}$$

$$\{\dot{\varepsilon}_{gm}\} = \frac{C_{Kg}}{3}\{\dot{\sigma}_m\} + \frac{1}{3\eta_{Kg}}\{\sigma_m\} \tag{D-2}$$

where, $C_{Gg}=1/G_g$, $C_{Kg}=1/K_g$, $\{\sigma'\}$: deviatoric stress, $\{\sigma_m\}$: mean normal
stress, G_g: shear modulus, K_g: bulk modulus.
Next, on Voigt model, representing generalized element by the subscript
i, we obtain

$$\{\sigma'\}=\frac{2}{C_{Gi}}\{\varepsilon_i{}'\}+2\eta_{Gi}\{\dot{\varepsilon}_i{}'\}, \qquad \{\sigma_m\}=\frac{3}{C_{Ki}}\{\varepsilon_{mi}\}+3\eta_{Ki}\{\dot{\varepsilon}_{mi}\}$$

Therefore,

$$\{\dot{\varepsilon}_i{}'\}=\frac{1}{2\eta_{Gi}}\{\sigma'\}-\frac{1}{T_{Gi}}\{\varepsilon_i{}'\} \tag{D-3}$$

$$\{\dot{\varepsilon}_{mi}\}=\frac{1}{3\eta_{Ki}}\{\sigma_m\}-\frac{1}{T_{Ki}}\{\varepsilon_{mi}\} \tag{D-4}$$

However, T_{Gi} and T_{Ki} show the retardation time of shear deformation and
volumetric deformation, respectively,

$$T_{Gi}=\eta_{Gi}C_{Gi}, \qquad T_{Ki}=\eta_{Ki}C_{Ki} \tag{D-5}$$

On generalized 4-element model, the stress is common to all the elements,
and the strain is the sum of each element. Since, from Eq.(D-1), (D-2),
(D-3) and (D-4), deviatoric component $\{\dot{\varepsilon}'\}$ and volumetric component
$\{\dot{\varepsilon}_m\}$ of the increment of total strain are given by

$$\{\dot{\varepsilon}'\}=\frac{C_{Gg}}{2}\{\dot{\sigma}'\}+(\frac{1}{2\eta_{Gg}}+\frac{1}{2\eta_{Gi}})\{\sigma'\}-\frac{1}{T_{Gi}}\{\varepsilon_i{}'\} \tag{D-6}$$

$$\{\dot{\varepsilon}_m\}=\frac{C_{Kg}}{3}\{\dot{\sigma}_m\}+(\frac{1}{3\eta_{Kg}}+\frac{1}{3\eta_{Ki}})\{\sigma_m\}-\frac{1}{T_{Ki}}\{\varepsilon_{mi}\} \tag{D-7}$$

Solving Eq.(D-6) and (D-7) on the rate of stress,

$$\{\dot{\sigma}'\}=\frac{2}{C_{Gg}}\{\dot{\varepsilon}'\}-\frac{1}{C_{Gg}}(\frac{1}{\eta_{Gg}}+\frac{1}{\eta_{Gi}})\{\sigma'\}+\frac{2}{C_{Gg}}\frac{1}{T_{Gi}}\{\varepsilon_i{}'\}$$

$$\{\dot{\sigma}_m\}=\frac{3}{C_{Kg}}\{\dot{\varepsilon}_m\}-\frac{1}{C_{Kg}}(\frac{1}{\eta_{Kg}}+\frac{1}{\eta_{Ki}})\{\sigma_m\}+\frac{3}{C_{Kg}}\frac{1}{T_{Ki}}\{\varepsilon_{mi}\} \tag{D-8}$$

$\{\sigma'\}$ and $\{\varepsilon'\}$ in Eq.(D-8) must be remarked to express tensor component.
From Eq.(D-8),

$$\{\dot{\sigma}\}=\{\dot{\sigma}'\}+\{\dot{\sigma}_m\}=\begin{Bmatrix} \dot{\sigma}_x \\ \dot{\sigma}_y \\ \dot{\sigma}_z \\ \dot{\tau}_{yz} \\ \dot{\tau}_{zx} \\ \dot{\tau}_{xy} \end{Bmatrix}$$

$$=\begin{bmatrix} \dfrac{1}{C_{Kg}}+\dfrac{4}{3C_{Gg}} & & & & & \text{sym.} \\ \dfrac{1}{C_{Kg}}-\dfrac{2}{3C_{Gg}} & \dfrac{1}{C_{Kg}}+\dfrac{4}{3C_{Gg}} & & & & \\ \dfrac{1}{C_{Kg}}-\dfrac{2}{3C_{Gg}} & \dfrac{1}{C_{Kg}}-\dfrac{2}{3C_{Gg}} & \dfrac{1}{C_{Kg}}+\dfrac{4}{3C_{Gg}} & & & \\ 0 & 0 & 0 & \dfrac{1}{C_{Gg}} & & \\ 0 & 0 & 0 & 0 & \dfrac{1}{C_{Gg}} & \\ 0 & 0 & 0 & 0 & 0 & \dfrac{1}{C_{Gg}} \end{bmatrix}\begin{Bmatrix} \dot{\varepsilon}_x \\ \dot{\varepsilon}_y \\ \dot{\varepsilon}_z \\ \dot{\gamma}_{yz} \\ \dot{\gamma}_{zx} \\ \dot{\gamma}_{xy} \end{Bmatrix}$$

$$+\frac{2}{C_{Gg}}\frac{1}{T_{Gi}}\begin{Bmatrix} \dot{\varepsilon}_x \\ \dot{\varepsilon}_y \\ \dot{\varepsilon}_z \\ \frac{1}{2}\gamma_{yz} \\ \frac{1}{2}\gamma_{zx} \\ \frac{1}{2}\gamma_{xy} \end{Bmatrix}i+\frac{3}{C_{Kg}}\frac{1}{T_{Ki}}\begin{Bmatrix} \varepsilon_m \\ \varepsilon_m \\ \varepsilon_m \\ 0 \\ 0 \\ 0 \end{Bmatrix}i-[P]\begin{Bmatrix} \sigma_x \\ \sigma_y \\ \sigma_z \\ \tau_{yz} \\ \tau_{zx} \\ \tau_{xy} \end{Bmatrix} \qquad \text{(D-9)}$$

However, the matrix expressed by $[P]$ and its component is

$$[P]=\frac{1}{3}\begin{bmatrix} A_K+2A_G & & & & & \text{sym.} \\ A_K-A_G & A_K+2A_G & & & & \\ A_K-A_G & A_K-A_G & A_K+2A_G & & & \\ 0 & 0 & 0 & 3A_G & & \\ 0 & 0 & 0 & 0 & 3A_G & \\ 0 & 0 & 0 & 0 & 0 & 3A_G \end{bmatrix} \qquad \text{(D-10)}$$

$$A_G=\frac{1}{T_{Gg}}+\frac{1}{C_{Gg}}\frac{1}{\eta_{Gi}}\ , \qquad A_K=\frac{1}{T_{Kg}}+\frac{1}{C_{Kg}}\frac{1}{\eta_{Ki}}\ ,$$

$$T_{Gg} = C_{Gg} \eta_{Gg} , \qquad T_{Kg} = C_{Kg} \eta_{Kg}$$

In Eq.(D-9), shear strain is followed by the definition of engineering. Eq.(D-9) is equivalent to Eq.(3) in our report.
From Eq.(D-9), Eq.(4) can be rewritten as

$$\{\dot{A}\} = -\frac{2}{C_{Gg}} \frac{1}{T_{Gi}} \{\dot{\varepsilon}_i\} - \frac{3}{C_{Kg}} \frac{1}{T_{Ki}} \{\varepsilon_{\blacksquare i}\} + [P]\{\sigma\} \qquad (D-11)$$

In general, the viscus matrix in Eq.(4) is given in the form

$$[\eta] = \begin{bmatrix} \eta_K + \frac{4}{3}\eta_G & \eta_K - \frac{2}{3}\eta_G & \eta_K - \frac{2}{3}\eta_G & 0 & 0 & 0 \\ & \eta_K + \frac{4}{3}\eta_G & \eta_K - \frac{2}{3}\eta_G & 0 & 0 & 0 \\ & & \eta_K + \frac{4}{3}\eta_G & 0 & 0 & 0 \\ & & & \eta_G & 0 & 0 \\ & & & & \eta_G & 0 \\ & \text{sym.} & & & & \eta_G \end{bmatrix}$$

Now, we will describe determinating method of rheological constants under three axial stress. In this report, we assume isotropic materials and Poisson's ratio ν_g and ν_i corresponded to two springs in Fig.11.
Although constitutive equation of viscoelastic 4-element model under generalized three axial stress is expressed by Eq.(D-9), we are trying to apply Eq.(D-9) to creep test under uni-axial loading, i.e. constant stress acts as $\sigma_x = -\sigma$. Here, except for the unrelated component on σ_x, ε_x, ε_y and ε_z, we obtain

$$\begin{Bmatrix} \dot{\sigma}_x \\ \dot{\sigma}_y \\ \dot{\sigma}_z \end{Bmatrix} = \begin{Bmatrix} \dot{\sigma}_x \\ 0 \\ 0 \end{Bmatrix} = \begin{bmatrix} \frac{1}{C_{Kg}} + \frac{4}{3C_{Gg}} & \frac{1}{C_{Kg}} - \frac{2}{3C_{Gg}} & \frac{1}{C_{Kg}} - \frac{2}{3C_{Gg}} \\ & \frac{1}{C_{Kg}} + \frac{4}{3C_{Gg}} & \frac{1}{C_{Kg}} - \frac{2}{3C_{Gg}} \\ \text{sym.} & & \frac{1}{C_{Kg}} + \frac{4}{3C_{Gg}} \end{bmatrix} \begin{Bmatrix} \dot{\varepsilon}_x \\ \dot{\varepsilon}_y \\ \dot{\varepsilon}_z \end{Bmatrix} + \frac{2}{C_{Gg}} \frac{1}{T_{Gi}} \begin{Bmatrix} \varepsilon_x \\ \varepsilon_y \\ \varepsilon_z \end{Bmatrix} i + \frac{3}{C_{Kg}} \frac{1}{T_{Ki}} \begin{Bmatrix} \varepsilon_{\blacksquare} \\ \varepsilon_{\blacksquare} \\ \varepsilon_{\blacksquare} \end{Bmatrix} i$$

$$-\frac{1}{3} \begin{bmatrix} A_K + 2A_G & A_K - A_G & A_K - A_G \\ & A_K + 2A_G & A_K - A_G \\ \text{sym.} & & A_K + 2A_G \end{bmatrix} \begin{Bmatrix} \sigma_x \\ 0 \\ 0 \end{Bmatrix} \qquad (D-12)$$

On the other hand, the constitutive equation of 4-element model under uni-axial stress is represented by

$$\dot{\sigma} = \frac{\dot{\varepsilon}}{C} + \frac{\varepsilon_I}{CT_I} - \frac{1}{C}\left(\frac{1}{\eta} + \frac{C_I}{T_I}\right)\sigma \qquad (D-13)$$

where, $C = 1/E$, $T_I = C_I \eta_I$

The relations of Eq. (6)~(11) in our report can be led to by determining rheological constants as is equivalent to Eq. (D-12) and (D-13). Consequently, Eq. (6)~(11) are confirmed by the argument mentioned above. It is clear that the relations presented by Dr. A. Tereutyev can also be expressed with our method, through a detailed investigation.

DISCUSSION CONTRIBUTIONS ON
KEATING AND HANNANT (page 137)

C. LEGRAND
INSA-UPS, Toulouse, France

S.A. JEFFERIS
Queen Mary and Westfield College, London, UK

I used a vaned bob about twenty years ago to measure
rheological properties of fresh mortars at rest and also
under vibration and I remember presenting some results
during the RILEM Symposium on Fresh Concrete at Leeds in
1973. I agree completely with the authors' conclusion,
particularly with the measurement of the initial yield value
which is a real physical characteristic and perhaps the only
one which we can obtain easily with this kind of material.

C. LEGRAND

The paper presents data on the time to failure and the
rotational speed of the vane. Have the authors analysed
these data to see if there is a constant angle of rotation
at failure? Many materials show a relatively constant
strain at failure over a considerable range of strain rates
provided that extreme rates are avoided. Clearly it is
difficult to deduce an unambiguous failure strain from vane
tests because of end effects etc. However, the angle of
rotation at failure will give an indication of whether the
material is relatively brittle or relatively plastic.
    For cement slurries, drilling fluids etc, the strain at
failure may have a significant influence on the start up
pressures following an interruption in pumping. The
pipework and the fluid will always be slightly compressible
and thus with a brittle gel there is potential for
progressive failure as the pressure wave propagates aling
the pipe. Thus for long flexible pipes or gassy fluids the
start up pressure may be a fraction of that which would be

obtained from a calculation based on yield stress, pipe length and pipe diameter. With plastic gels where the peak and residual (after failure) strengths are quite similar the start up pressure should be higher than for a brittle gel.

The paper also discusses the influence of confining pressure on the vane strength and suggests that the observed increase in strength may be due to the change in effective stress between the cement particles. If this is the case then the effect should increase with time as water is taken up by hydration. Unfortunately, as the authors explain, the rapid rise in strength makes it impossible to examine the influence of pressure at later times. However, at the pressure used in the consistometer it would seem that the compressibility effects would have some influence and thus should be investigated as well as effective stress effects.

S.A. JEFFERIS

AUTHORS' REPLY (J. KEATING AND D.J. HANNANT)

We would like to thank Drs. Legrand and Jefferis for their valuable comments. In reply to the latter's query relating to the strain at failure at different rotation rates, an absolute answer cannot be given because the thickness of the sheared layer is not known. However, the time to peak torque varied approximately in proportion to the rotation rate with faster rotation giving shorter times, implying that the angle of rotation at peak torque was approximately constant between 0.06 and 6 rev/min. We would agree that the effect of confining pressure should be more apparent at later ages when more water has been taken up by hydration. However, the rapid increases in gel strength after 60 minutes made it impossible to establish experimentally whether this was the case. The compressibility effects due to the residual air content and compressibility of the solids could also have altered the effective stress and it would be very difficult to separately quantify these two effects.

DISCUSSION CONTRIBUTION ON MUSZYNSKI AND MIERZWA
(page 201)

S.A. JEFFERIS
Queen Mary and Westfield College, London, UK

I was very interested in this paper as we have found exactly
the same problems in the U.K. It is not common practice to
include grout rheology as a control parameter in cement
manufacture and thus supplies from different plants may
behave quite differently when used in prestressing ducts.
Thus when grouting two approaches are possible: (i) to
adjust the water/cement ratio to obtain suitable properties
or (ii) when possible, to use cement from a single proven
source. Except when working on jobs in a limited
geographical region it will usually be necessary to adopt
approach (i).
Typically what water/cement ratio is used for ducts in
Poland and what range of water/cement ratios does the
mathematical analysis show when applied to the range of
Polish cements? Have the authors found that there are
cements which the analysis would suggest are unsuitable for
grouting work?

AUTHORS' REPLY (W. MUSZYNSKI AND J. MIERZWA)

We are very glad to hear that Dr. Jefferis has found exactly
the same problems in the U.K. as we have described in our
paper. Generally speaking we agree that two approaches are
possible when grouting. However, the results of our
investigations have proved that high quality grouting
depends to a great extent on using the appropriate cement.
The water/cement ratio used in Poland for ducts is from 0.38
to 0.48. According to our national standards, Portland
cements of class 35 are applicable for grouting and it is
required that the amount of hydraulic additives in a
production process shall not exceed 10%.
Applying the mathematical analysis to the range of Polish
cements predicts water/cement ratios from 0.38 to 0.48.
This is for Polish cements of Blaine fineness between 280
and 350 m^2/kg and of M_{SA} modulus between 3.5 to 5.3, where

$$M_{SA} = (C_3A + C_2S)/(C_3A + C_4AF).$$

We have found that there are cements determined by their
basic properties such as appropriate strength, M_{SA} modulus
and Blaine fineness which are suitable for grouting work and
any other cements are not recommended.

DISCUSSION CONTRIBUTION ON BORGESSON AND FREDRIKSSON
(page 313)

C. ELLIS
Sheffield City Polytechnic, UK
G.H. TATTERSALL
University of Sheffield, UK

Fig. 4 of the paper shows that the logarithmic plot of amplitude, A, against frequency, f, is a straight line of slope -3/2. If one makes the a priori assumption that power, W, is proportional to a function of amplitude and frequency then for dimensional compatibility the relationship could be presented at its simplest as follows:

$$W = kMA^2f^3$$

where k is a constant and M is a term representing mass. Power may be expressed in Nms^{-1} or fundamentally in kgm^2s^{-3} (SI units). Taking natural logarithms the expression becomes

$$lnW = ln(kM) + 2lnA + 3lnf.$$

Rearranging gives

$$lnA = (ln(W/kM))/2 - (3lnf)/2.$$

This confirmation of the relationship may be fortuitous but merits further examination.

C. ELLIS

This paper deserves detailed study for comparison of the results and conclusions with those of other workers, and I offer only the following for immediate comment.

The findings of Newtonian behaviour of the vibrated material and the lack of effectiveness of vibration at higher frequencies are in broad agreement with my own, but

Fig. 4 of the paper seems to show a disagreement on an important point. The slope of the line is -3/2 so the equation of the line is

$$\ln \gamma_a = \ln K - 3/2 \ln f$$

where K is a constant. This may be rewritten as

$$\gamma_a f^{3/2} = K$$

for a constant value of m (m = 1). In other words, the effectiveness of vibration depends on

amplitude x (frequency)$^{3/2}$.

This contrasts with my findings, for both cement pastes and concretes, that the important parameter is velocity, i.e. amplitude x frequency.

G.H. TATTERSALL

DISCUSSION CONTRIBUTION ON BARRIOULET AND LEGRAND
(page 343)

G.H. TATTERSALL
University of Sheffield, UK

Could the authors summarise what experimental evidence is available to show that their 'calcite filler mix' provides a satisfactory simulation of the behaviour of cement paste?

AUTHORS' REPLY (M. BARRIOULET AND C. LEGRAND)

For a long time we have published flow curves, both at rest and under vibration, of mixes of water with different mineral powders such as crushed calcite, quartz or feldspar as well as with various cements (Portland, slag etc.). Clearly the values of the characteristics vary but we have never noticed fundamental differences in the rheological behaviour of such mixes. We have studied real cement pastes using several workability tests and also the apparatus presented in our paper and have observed the same phenomena as with the mineral powder pastes. Therefore, qualitatively the simulation seems to be satisfactory and is very useful in practice because it allows us to work with a non-setting material and preserve the water whose special properties are very important in the rheology of pastes.

INDEX